目　錄

主題 A：NodeMCU 開發工具與驅動程式安裝

1. 題目：認識 NodeMCU 硬體平台，並且安裝相關的驅動程式以及下載並安裝整合開發環境。

2. 實驗目標：熟悉 NodeMCU 軟硬體平台。

3. 實驗步驟：此實驗將分四個階段進行，

   a. 驅動程式：USB-TTL Driver CH341 或是 CP210X。

   b. 韌體燒錄程式：NodeMCU Flasher(optional)。

   c. 整合開發工具：ESPlorer。

   d. 電路與 PCB 印刷電路板的設計工具：Fritzing。

a. 安裝驅動程式：USB-TTL Driver CH341 或是 CP210X。

   (1) 經由網頁 http://www.electrodragon.com/w/CH341 下載 USB-TTL Driver CH341，如圖 1-1；
   或是下載 USB-TTL Driver CP210X
   https://www.silabs.com/products/mcu/Pages/USBtoUARTBridgeVCPDrivers.aspx，如圖 1-2。

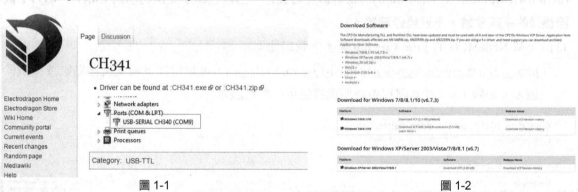

圖 1-1   圖 1-2

註：依據作業系統環境選擇紅框處要下載的版本。

   (2) 經由下載檔案位置安裝將 USB 轉成序列埠的 USB-TTL Driver，如圖 1-3。

圖 1-3

(3) 將電源線連接 NodeMCU，進入裝置管理員(以 Win 7 為例，在電腦按下右鍵選擇管理，如圖 1-4)可以看到連接埠已經有接上 NodeMCU 裝置，此時 NodeMCU 的連接埠為"COM5"，如圖 1-5。

圖 1-4　　　　　　　　　　　　　　　　圖 1-5

b. 安裝韌體燒錄程式：將 NodeMCU 即時作業系統的韌體(Firmware)燒錄到 NodeMCU 中，讓使用者可以用電腦和 NodeMCU 溝通。**(注意：如果是購買全華圖書或者是哥大智慧科技(http://www.pcstore.com.tw/aiot/)的物聯網實習套件，已經內建了支援物聯網 CoAP、MQTT 國際標準的系統軟體，不要燒錄韌體。)**

(1) 下載 NodeMCU 的 Firmware bin 檔，下載網址：
https://github.com/nodemcu/nodemcu-firmware/releases，可以選擇下載支援浮點數或整數的版本，如圖 1-6，在本課程中我們選擇使用浮點數版本。

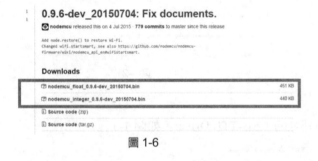

圖 1-6　　　　　　　　　　　　　　　　圖 1-7

(2) 下載要安裝 NodeMCU firmware 時要使用的執行檔，由網頁
https://github.com/nodemcu/nodemcu-flasher 下載 NodeMCU Flasher，經由圖 1-7 的第 1 和第 2 步驟將檔案下載至電腦。

(3) 將下載檔案解壓縮後，依照作業系統版本點選要執行的執行檔"ESP8266Flasher.exe"，如圖 1-8 和圖 1-9。

圖 1-8

圖 1-9

(4) 選擇 NodeMCU 的連接埠，如圖 1-10 選擇 COM5。

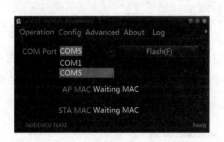

圖 1-10

圖 1-11

(5) 點選上方 Config，再選擇圖 1-11 的紅框按鍵，準備將 Flasher 的 bin 檔載入。

(6) 選擇下載 bin 檔的存放位置，如圖 1-12。

圖 1-12

(7) 燒錄位置選擇"0x00000"，如圖 1-13。

圖 1-13

(8) Advanced 的 Baudrate 可重設爲 9600 或維持 115200 皆可，如圖 1-14。

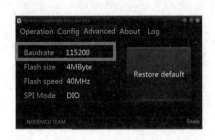

圖 1-14　　　　　　　　　　　　　　　　　圖 1-15

(9) 返回 Operation 頁面，點選 Flash(F)按鈕，開始燒錄 Firmware。燒錄過程會有藍色的進度顯示，當左下方出現綠色勾勾表示燒錄完成，如圖 1-15。

c. 安裝整合開發工具：ESPlorer。

(1) 軟體下載網址：https://drive.google.com/file/d/1KIFLtANy8LsLrUjwh5bO_dhjcg7Uyuh6/view or http://esp8266.ru/esplorer/，將網頁下拉至下方 ESPlorer Downloads 選擇檔案下載，如圖 1-16。

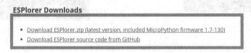

圖 1-16

(2) 將 ESPlorer.zip 解壓後，執行 ESPlorer.bat，如圖 1-17。開啓後即可進入開發環境 IDE，如圖 1-18。目前最新的 ESPlorer IDE 版本爲 v0.2.0-rc3。

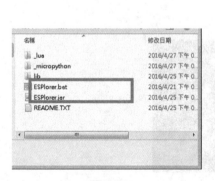
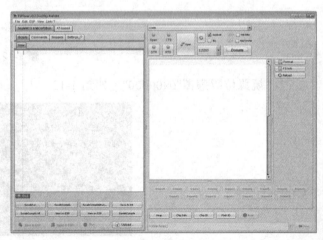

圖 1-17　　　　　　　　　　　　　　　　　圖 1-18

d. 安裝電路與 PCB 印刷電路板的設計工具：Fritzing。Fritzing 是一種開放原始碼的工具，可利用 Fritzing 來製作嵌入式系統電路設計。

(1) 軟體下載網址：http://fritzing.org，點擊 圖 1-19 中紅框處進行下載。

(2) 選擇"No Donation"後點擊"Download"， 如圖 1-20。接著選擇作業系統版本下 載，如圖 1-21。

(3) Fritzing 是免安裝版本，直接點選 Fritzing.exe 即可開啓工具開始編輯，如 圖 1-22。

圖 1-19

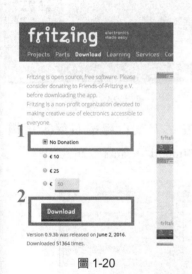

圖 1-20

圖 1-21

圖 1-22

(4) Fritzing 執行環境如圖 1-23。

圖 1-23

主題 B：使用 Fritzing IDE 平台做 LED Blinking 電路設計

1. 題目：使用 Fritzing 來製作讓 LED 燈閃爍的電路設計，其中 LED 燈接腳使用 NodeMCU 的 GPIO D1。

2. 實驗目標：使用 Blink 範例練習 Fritzing 的電路設計，實際產生麵包板電路、電路概要圖以及 PCB 印刷板電路。

3. 實驗步驟：此實驗將分三個階段進行，

   a.　產生麵包板電路。

   b.　產生電路概要圖。

   c.　產生 PCB 印刷板電路圖。

a.　產生麵包板電路

   (1)　開啓 Fritzing 工具，點選"麵包板"，如圖 1-24。

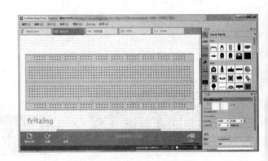

圖 1-24

   (2)　目前的 Fritzing 版本並沒有內建 NodeMCU，因此可以透過網址下載：

   https://github.com/squix78/esp8266-fritzing-parts

   ，如圖 1-25。

圖 1-25

   (3)　在右方空白處按右鍵，選擇 "Import"，如圖 1-26。

圖 1-26

6

(4)　選擇在第(2)步中所下載的檔案位置下的 nodemcu-v1.0 子目錄中的 NodeMCUV1.0.fzpz，如圖 1-27。

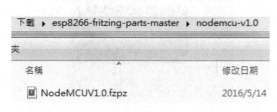

圖 1-27

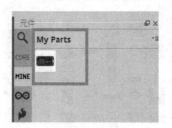

圖 1-28

(5)　之後可以藉由搜尋元件 NodeMCU 即可，如圖 1-28。

註：麵包板藍(黑)邊是 GND，紅邊是電源輸入。

(6)　將 NodeMCU 往麵包板上拖曳並放置在麵包板上，如圖 1-29。放置 NodeMCU 時需注意上下要留接孔以便拉線。

圖 1-29

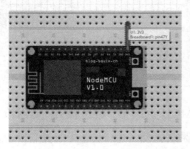

圖 1-30

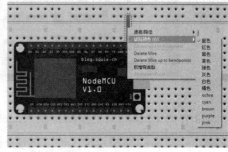

圖 1-31

(7)　將 NodeMCU 的 3V3 接到麵包板上的紅邊，GND 接到藍邊。拉線方式為按下接腳的接孔拖拉至麵包板上的接孔，如圖 1-30。如果要變更接線線條的顏色，可點選線條按下右鍵，選擇"線路顏色"變更，如圖 1-31。拉線完成圖如圖 1-32。

圖 1-32

(8) 如圖 1-33 搜尋"LED"，將想要的 LED 拖曳至麵包板上，如圖 1-34。

圖 1-33

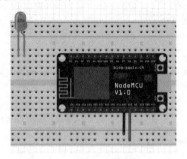

圖 1-34

(9) 如圖 1-35 搜尋"resistor"，將電阻拖曳至麵包板上，並將電阻的一邊和 LED 的短邊接腳相接，如圖 1-36。

圖 1-35

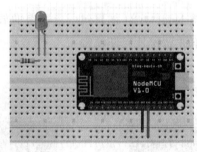

圖 1-36

(10) 將 LED 的長邊接腳接到 NodeMCU 上的 D1 數位接孔，電阻沒有和 LED 短邊接腳相接的那一端接到麵包板上的 GND。線路盡量不要跨過其他元件，因此可以拉住線條改變線路走向，如圖 1-37。

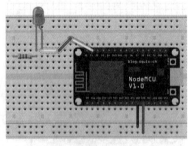

圖 1-37

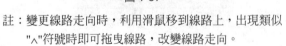

註：變更線路走向時，利用滑鼠移到線路上，出現類似"∧"符號時即可拖曳線路，改變線路走向。

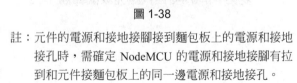

圖 1-38

註：元件的電源和接地接腳接到麵包板上的電源和接地接孔時，需確定 NodeMCU 的電源和接地接腳有拉到和元件接麵包板上的同一邊電源和接地接孔。

(11) 完成接線圖如圖 1-38。

b. 產生電路概要圖

(1) 使用 Fritzing 建構好麵包板電路後，點選上方的"概要圖"標籤即可看到自動幫我們建好的電路概要圖，如圖 1-39。

(2) 使用者可以對每個元件做旋轉或更改放置位置以及修改線路走向來完成美觀的電路概要圖。元件擺放位置可以利用拖拉的方式移動，而元件方向也可以利用下方的旋轉按鈕轉動，如圖 1-40。以圖 1-41 為例，可以變更元件的擺放位置並視需要旋轉元件。

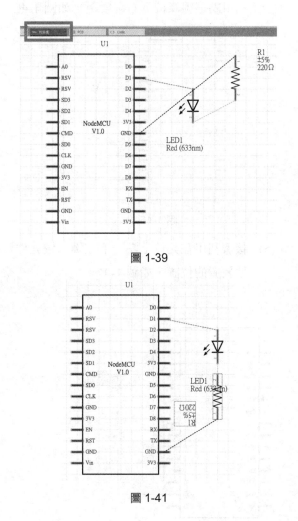

圖 1-39

圖 1-40

(3) 使用者可以對每個元件變更命名，例如要將 NodeMCU 的標題 "U1" 改為 "NodeMCU"。

    (a) 點選標題"U1"按右鍵，如圖 1-42。

    (b) 選擇編輯後，出現設定標頭文字的視窗，輸入"NodeMCU"，如圖 1-43。

圖 1-41

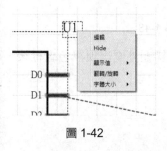

圖 1-42

圖 1-43

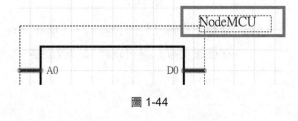

圖 1-44

    (c) 按下 OK 後，標題文字就會改為 NodeMCU 了，如圖 1-44。

(4) 線路變更方式如同麵包板的作法。以圖 1-45 為例，可以變更線路的走向。線路顏色也可以利用點選線條按下右鍵變更，如圖 1-46。

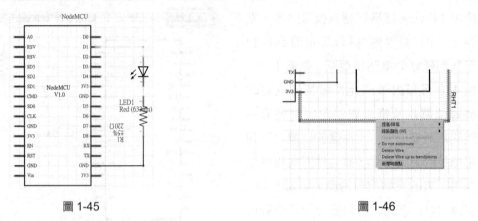

圖 1-45　　　　　　　　　　　　　　　　　圖 1-46

(5) 接著利用在元件上點一下左鍵，讓元件及標題外圍出現黑色虛線後，再點著標題移動到想要放置的位置，如圖 1-47。

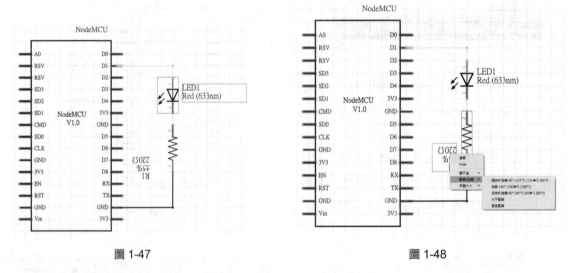

圖 1-47　　　　　　　　　　　　　　　　　圖 1-48

(6) 當想要旋轉元件的標題時，一樣在元件上點一下左鍵，讓元件及標題外圍出現黑色虛線後，在標題上按右鍵再選擇"翻轉/旋轉"，如圖 1-48。

c. 產生 PCB 印刷板電路圖

(1) 使用 Fritzing 建構好麵包板電路後，點選上方的"PCB"標籤即可看到自動幫我們建好的 PCB 印刷板電路圖，如圖 1-49。

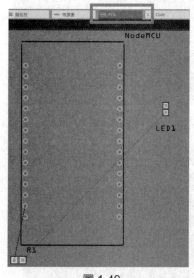

圖 1-49

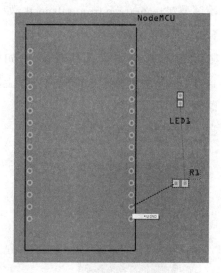

圖 1-50

註：移動 R1 位置後可以發現，會自動將 GND 接腳移動 到右方，讓線路不跨過 NodeMCU 元件。

(2) 轉好的 PCB 圖或許不美觀或是線的接孔不對，我們可以進一步修正。首先移動 R1 至想要 的位置，如圖 1-50。

(3) 讓 R1 轉向，使得之後線路連接可以更 美觀，旋轉方式如同電路概要圖的作 法。而轉向後由於 R1 的標題會跟著旋 轉，因此可以利用點選標題按右鍵，選 擇"翻轉/旋轉"調整標題的方向，如圖 1-51。

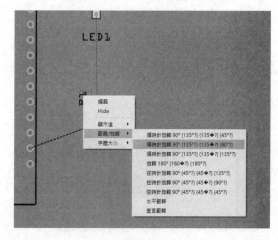

圖 1-51

(4) 若要變更線路走向，則如同電路概要圖
的作法一樣。當線路是虛線狀態時，表
示還沒有佈線完成，因此點選線條按右
鍵選擇"Create trace from ratsnest"，如圖
1-52。

(5) 點選後該條線路就佈線完成，如圖
1-53；另一種佈線方式則是在虛線上點
擊兩次也可以。

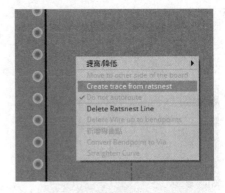

圖 1-52

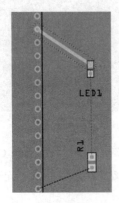

圖 1-53

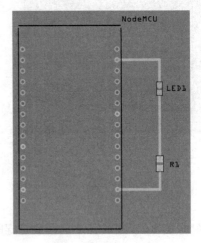

圖 1-54

(6) 將剩下的虛線線路完成佈線如圖 1-54。

4. 實驗成果：

(1) 麵包板電路如圖 1-55。

(2) 電路概要圖如圖 1-56。

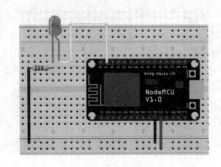

圖 1-55

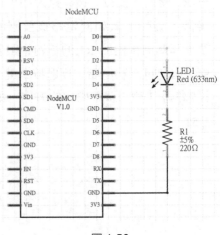

圖 1-56

12

(3) PCB 印刷板電路圖如圖 1-57。

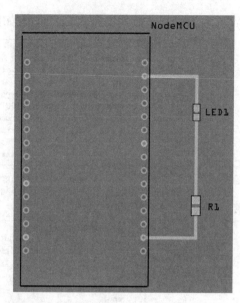

圖 1-57

主題 C：使用 Snap4NodeMCU 與 ESPlorer IDE 平台做 LED Blinking 程式設計

1. 題目：使用 ESPlorer IDE 平台撰寫讓 LED 燈閃爍的 Lua 程式，其中 LED 燈接腳使用 NodeMCU 的 GPIO D1，並實際利用麵包板將 NodeMCU 以及 LED 進行線路連接。

2. 實驗目標：使用 Blink 範例程式練習 Lua 的程式設計，並實際觀察麵包板接線完成後，程式進行時 ESPlorer 終端機顯示畫面以及麵包板上 LED 閃爍狀況。

3. 實驗步驟：

(1) 點擊 ESPlorer.bat 工具，進入開發環境 IDE，如圖 1-58。

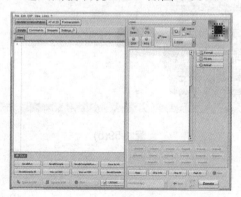

圖 1-58

13

(2) 在左方的程式撰寫區撰寫 Lua 程式，如圖 1-59(a)。也可以使用 Snap4NodeMCU，如圖 1-59(b)。

```lua
1   --defined pin 1 witch will used
2   pin = 1
3   --defined state of the pin 1
4   state = "off"
5   --defined pin 1 become gpio output
6   gpio.mode(pin,gpio.OUTPUT)
7   --1 second alarm the process 0 and do one time
8   tmr.alarm(0,1000,1,function()
9       --if now is "close" open the led and change state
10      if state == "off" then
11          --make pin output high
12          gpio.write(pin,gpio.HIGH)
13          print("GPIO 1 light on")
14          state = "on"
15      --if now is "open" close the led and change state
16      else
17          --make pin output low
18          gpio.write(pin,gpio.LOW)
19          print("GPIO 1 light off")
20          state ="off"
21      end
22  end)
```

圖 1-59(a)

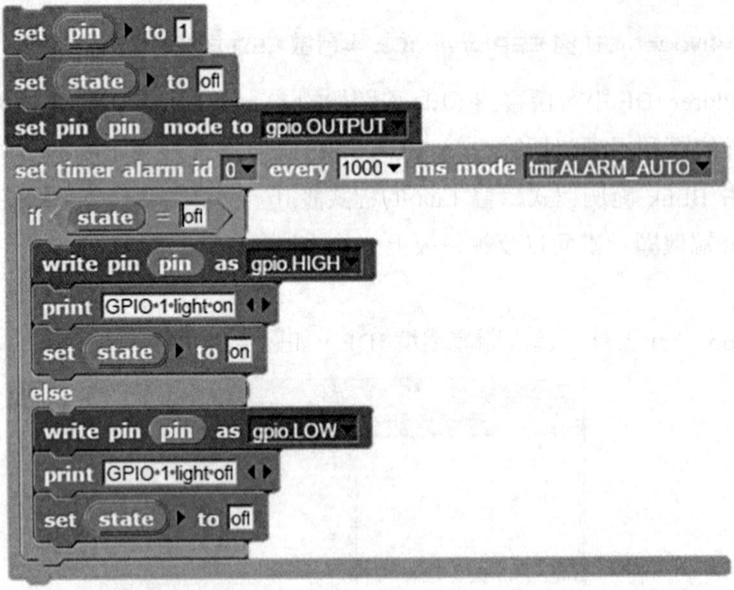

圖 1-59(b)

(3) 將 NodeMCU 插入到麵包板上，如圖 1-60。

(4) 將 NodeMCU 的電源(3V3)接腳利用杜邦線接到麵包板上的電源接孔，如圖 1-61。

(5) 將 NodeMCU 的接地(GND)接腳利用杜邦線接到麵包板上的接地接孔，如圖 1-62。

圖 1-60

圖 1-61

圖 1-62

(6) 將 LED 燈插入到麵包板上，如圖 1-63。

(7) 將電阻插入麵包板中，其中一端和 LED 燈的短邊接腳相接，如圖 1-64。

(8) 將 LED 燈的長邊接腳接到 NodeMCU 的 D1 接腳，如圖 1-65。

圖 1-63

圖 1-64

圖 1-65

(9) 將電阻另一端不是和 LED 燈的短邊接腳相接的接腳接到麵包板上的接地接孔，如圖 1-66。

(10) 確定線路沒問題後，利用 MicroUSB-USB 接線，MicroUSB 端接上 NodeMCU，USB 接到電腦的 USB 插孔，如圖 1-67。

圖 1-66

圖 1-67

(11) 在 ESPlorer 右方視窗按下"Reflash"按鈕，選擇連接埠"COM5"，如圖 1-68。

圖 1-68

(12) 接著點擊"Open"按鈕(圖 1-69)，建立 NodeMCU 與 ESPlorer 之間的連線，如圖 1-70。

圖 1-69

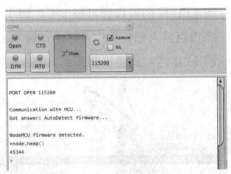
圖 1-70

(13) 點擊左下方"Save to ESP"按鈕，如圖 1-71。

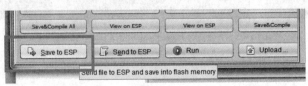
圖 1-71

(14) 點擊後跳出儲存視窗，可以變更檔案名稱以及儲存位置後進行存檔，如圖 1-72。

圖 1-72 　　　　　　　　　　　　　　　　　　圖 1-73

(15) 存檔後同時會將程式一行一行的寫入到 NodeMCU 中，如圖 1-73 的終端機畫面。

4. 實驗成果：

(1) ESPlorer 終端機畫面，如圖 1-74。

(2) 麵包板上的 LED 燈也會跟著閃爍如圖 1-75。

```
-
GPIO 1 light on
GPIO 1 light off
GPIO 1 light on
GPIO 1 light off
GPIO 1 light on
GPIO 1 light off
GPIO 1 light on
GPIO 1 light off
GPIO 1 light on
GPIO 1 light off
GPIO 1 light on
GPIO 1 light off
GPIO 1 light on
GPIO 1 light off
GPIO 1 light on
GPIO 1 light off
GPIO 1 light on
GPIO 1 light off
GPIO 1 light on
GPIO 1 light off
```

圖 1-74 　　　　　　　　　　　　　　　　　　圖 1-75

主題 D：使用 Snap4NodeMCU 與 ESPlorer IDE 平台做 Wi-Fi 連線程式設計

1. 題目：使用 ESPlorer IDE 平台撰寫啟動 Wi-Fi 連線並取得 IP Address。

2. 實驗目標：使用 Wi-Fi 連線範例程式練習 Lua 的程式設計，並實際觀察程式進行時 ESPlorer 終端機顯示畫面。

3. 實驗步驟：

    (1) 點擊 ESPlorer.bat 工具，進入開發環境 IDE，如圖 1-76。

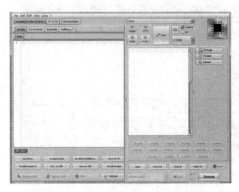

圖 1-76

    (2) 在左方的程式撰寫區撰寫 Lua 程式，如圖 1-77(a)，其中 4Clab-2.4G 為無線路由器的 SSID，而 socad3284 為路由器的密碼，這兩個值必須改成所在環境路由器的 SSID 和密碼。也可以使用 Snap4NodeMCU，如圖 1-77(b)。

```
1   print("Connect to AP:")
2   wifi.setmode(wifi.STATION)
3   wifi.sta.config("4Clab-2.4G", "socad3284")
4   tmr.alarm(0,1000,1, function()
5     print(wifi.sta.getip())
6     if wifi.sta.getip()~=nil then
7         tmr.stop(0)
8     end
9   end)
```

圖 1-77(a)

```
set Wi-Fi mode wifi.STATION ▼
Wi-Fi connect SSID  4Clab-2.4G  as string  password socad3284
set timer alarm id 0 ▼ every 1000 ▼ ms mode tmr.ALARM_AUTO ▼
get IP address,netmask,gateway to  ip    netmask    gateway
print  ip=  ip  ◀▶
if  not  ip  =  nil  ▶
tmr.stop(): stop timer 0 ▼
```

圖 1-77(b)

18

4. 實驗成果：

終端機顯示結果畫面如圖 1-78：NodeMCU 所獲得的的 IP 位址是 192.168.1.212，而 AP 的 IP 位址是 192.168.1.1。

```
> file.remove("wifi_ch1-2.lua");
> file.open("wifi_ch1-2.lua","w+");
> w = file.writeline
> w([[print("Connect to AP:")]]);
> w([[wifi.setmode(wifi.STATION)]]);
> w([[wifi.sta.config("4Clab-2.4G", "socad3284")]]);
> w([[tmr.alarm(0,1000,1, function()]]);
> w([[   print(wifi.sta.getip())]]);
> w([[   if wifi.sta.getip()~=nil then]]);
> w([[      tmr.stop(0)]]);
> w([[  end]]);
> w([[end)]]);
> file.close();
> dofile("wifi_ch1-2.lua");
Connect to AP:
> nil
nil
nil
nil
nil
192.168.1.212   255.255.255.0   192.168.1.1
```

圖 1-78

19

主題 A：使用 Fritzing IDE 平台做 LED Dimming 電路設計

1. 題目：使用 Fritzing 來製作讓 LED Dimming 的電路設計，其中 LED 燈接腳使用 NodeMCU 的 GPIO D3。
2. 實驗目標：使用 Dimming 範例練習 Fritzing 的電路設計，實際產生麵包板電路、電路概要圖以 及 PCB 印刷板電路。
3. 實驗步驟：此實驗將分三個階段進行，
   a. 產生麵包板電路圖。
   b. 產生電路概要圖。
   c. 產生 PCB 印刷板電路圖。

a. 產生麵包板電路圖
   (1) 開啓 Fritzing 工具，點選"麵包板"，如圖 2-1。

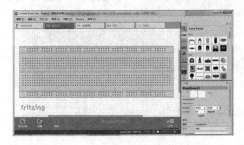 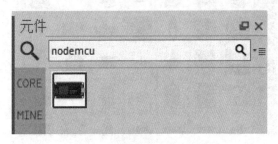

圖 2-1 圖 2-2

   (2) 在右方搜尋窗格輸入"NodeMCU"，找到 NodeMCU 元件，如圖 2-2。
   (3) 將 NodeMCU 往麵包板上拖曳並放置在麵包板上，如圖 2-3。

圖 2-3 圖 2-4

   (4) 將 NodeMCU 的 3V3 接到麵包板上的紅邊，GND 接到藍邊，如圖 2-4。

(5) 如圖 2-5 搜尋"LED"，將想要的 LED 拖曳至麵包板上，如圖 2-6。

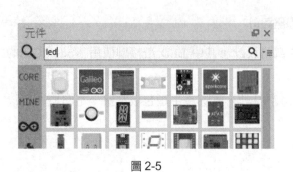

圖 2-5

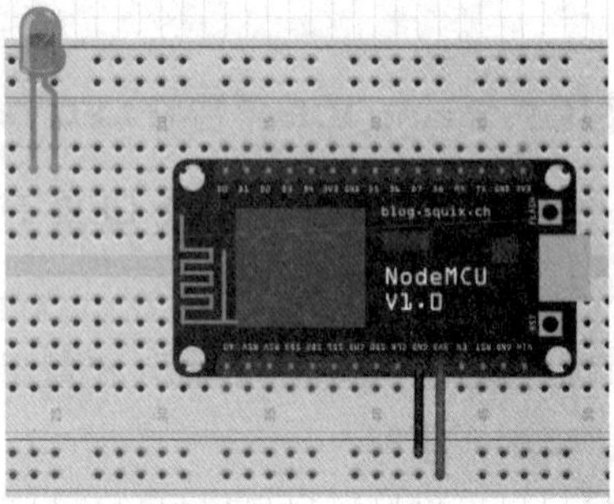

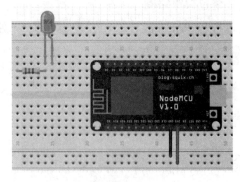

圖 2-6

(6) 搜尋"resistor"，將電阻拖曳至麵包板上一邊和 LED 的短邊接腳相接，如圖 2-7 和圖 2-8。

圖 2-7

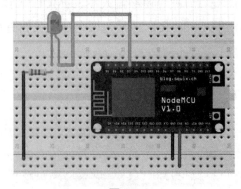

圖 2-8

(7) 將 LED 的長邊接腳接到麵包板上的 D3 數位接孔，電阻沒和 LED 相接的另一端接腳接到麵包板上的 GND，如圖 2-9。

圖 2-9

b. 產生電路概要圖

(1) 點選上方的"概要圖"標籤即可看到自動幫我們建好的電路概要圖，如圖 2-10。

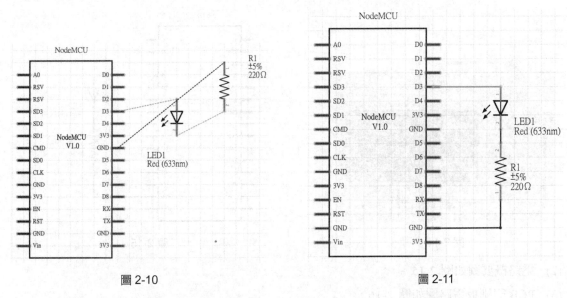

圖 2-10   圖 2-11

(2) 將元件以及線路擺放至較美觀的位置，並將線路佈線完成，如圖 2-11。

c. 產生 PCB 印刷板電路圖

(1) 點選上方的"PCB"標籤看到自動幫我們建好的 PCB 印刷板電路圖，如圖 2-12。

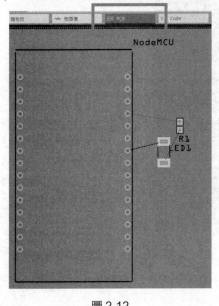 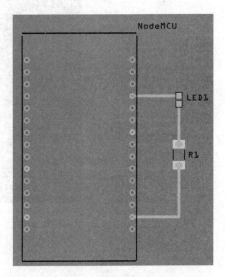

圖 2-12   圖 2-13

(2) 將元件以及線路擺放至較美觀的位置，並將線路佈線完成，如圖 2-13。

23

4. 實驗成果：

(1) 麵包板電路如圖 2-14。

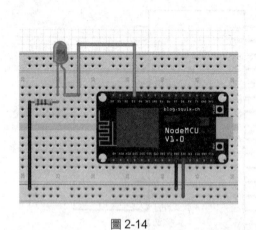

圖 2-14

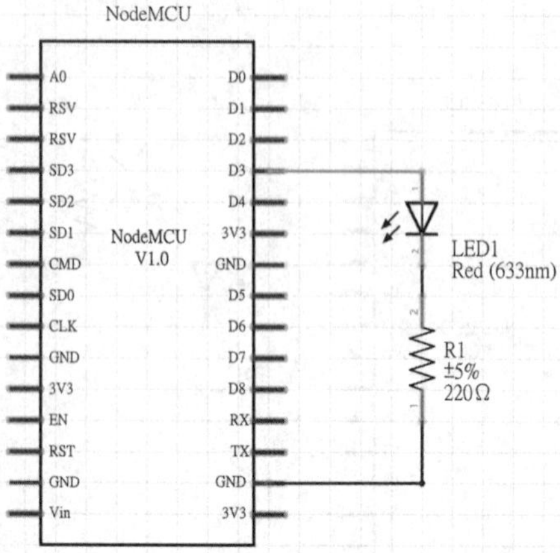

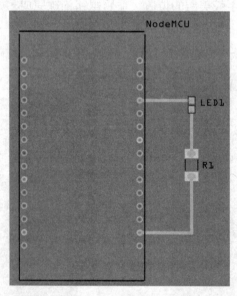

圖 2-15

(2) 電路概要圖如圖 2-15。

(3) PCB 印刷板電路圖如圖 2-16。

圖 2-16

主題 B：使用 Snap4NodeMCU 與 ESPlorer IDE 平台做 LED Dimming 程式設計

1. 題目：使用 ESPlorer IDE 平台撰寫讓 LED 燈 Dimming 的 Lua 程式，其中 LED 燈接腳使用 NodeMCU 的 GPIO D3，並實際利用麵包板將 NodeMCU 以及 LED 進行線路連接。
2. 實驗目標：使用 Dimming 範例程式練習 Lua 的程式設計，並實際觀察麵包板接線完成後，程式進行時 ESPlorer 終端機顯示畫面以及麵包板上 LED 閃爍狀況，讓使用者實際了解 PWM 運作情形。
3. 實驗步驟：此實驗將分三個階段進行，
   a. 讓 LED 燈顯示從最亮到最暗，如此重複不斷。
   b. 讓 LED 燈顯示從最亮到最暗，再從最暗到最亮，如此重複不斷。
   c. 讓 LED 燈閃爍顯示(即一秒亮一秒暗)從最亮到最暗，再從最暗到最亮，如此重複不斷。

a. 讓 LED 燈顯示從最亮到最暗，如此重複不斷
   (1) 點擊 ESPlorer.bat 工具，進入開發環境 IDE，如圖 2-17。

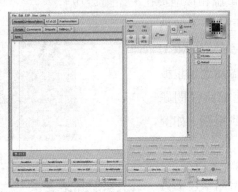

圖 2-17

   (2) 在左方的程式撰寫區撰寫 Lua 程式，利用每次 timer 的呼叫逐漸下降 PWM 輸出之平均電壓，使 LED 亮度逐漸降低到最暗時，重設 duty cycle 到最大值，如圖 2-18(a)。也可以使用 Snap4NodeMCU 讓 LED 燈顯示從最亮到最暗，如此重複不斷，如圖 2-18(b)。

```
1   pin=3                      -- set PWM to GPIO pin 3
2   dc=1023                    -- set duty cycle to 1023
3   gpio.mode(pin,gpio.OUTPUT) -- set pin 3 to be an OUTPUT pin
4   -- set clock to 1000HZ (period = 0.001 second)
5   pwm.setup(pin,1000,dc)     -- set pin 3 as PWM pin and
6   pwm.start(pin)             -- start PWM output
7   tmr.alarm(0,100,tmr.ALARM_AUTO,function() -- repeat every 0.1 second
8     print("dc=",dc)          -- debug dc
9     pwm.setduty(pin,dc)      -- change PWM duty cycle (dc)
10    if (dc < 20) then
11      dc=1023
12    else
13      dc=(dc - 20)           -- reduce dc by 20 every cycle
14    end
15  end)
```

圖 2-18(a)

25

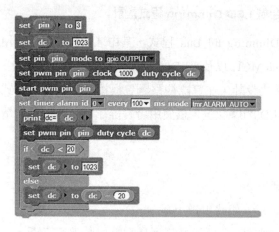

圖 2-18(b)

圖 2-19

(3) 將 NodeMCU 插入到麵包板上，再利用杜邦線把 NodeMCU 的電源(3V3)接腳和接地(GND)接腳接到麵包板上的電源和接地接孔，如圖 2-19。

(4) 將 LED 燈插入到麵包板上，並把 LED 燈的長邊接腳接到 NodeMCU 的 D3 接腳，短邊接腳接到麵包板上的接地接孔。確定線路沒問題後，將 NodeMCU 接到電腦，如圖 2-20。

圖 2-20

(5) 到 ESPlorer IDE 平台點擊左下方 "Save to ESP"按鈕，將程式存檔並寫入到 NodeMCU 後執行，如圖 2-21。

```
> w = file.writelinew([[--set PWM pin to gpio pin 3]]);
> w([[pin = 3]]);
> w([[--set pin 3 to be OUTPUT pin]]);
> w([[gpio.mode(pin,gpio.OUTPUT)]]);
> w([[--set the highest brightness]]);
> w([[width = 1023]]);
> w([[--set the clock to 1000Hz]]);
> w([[clock = 1000]]);
> w([[--set up pin 3 with 1000Hz]]);
> w([[pwm.setup(pin,clock,width)]]);
> w([[--start PWM output]]);
> w([[pwm.start(pin)]]);
> w([[--change the pulse width each 0.1 sec]]);
> w([[tmr.alarm(0,100,1,function()]]);
> w([[    if width < 20 then]]);
> w([[        width = 1023]]);
> w([[    else ]]);
> w([[        width = width -20]]);
> w([[    end]]);
>
```

圖 2-21

26

b. 讓 LED 燈顯示從最亮到最暗,再從最暗到最亮,如此重複不斷

(1) 修改步驟 a 的程式碼,利用每次 timer 呼叫使 duty cycle 逐漸減少,也就是逐漸下降 PWM 輸出之平均電壓,使 LED 亮度逐漸降低直到最暗時,再逐漸增加 duty cycle 的值,如圖 2-22(a)。也可以使用 Snap4NodeMCU 讓 LED 燈顯示從最亮到最暗,再從最暗到最亮,如此重複不斷,如圖 2-22(b)。

```
1   pin=3                                    -- set PWM to GPIO pin 3
2   dc=1023                                  -- set duty cycle to 1023
3   ascdec=-1                                -- set dc change ascending or descending
4   inc=20                                   -- change delta value
5   gpio.mode(pin,gpio.OUTPUT)               -- set pin 3 to be an OUTPUT pin
6   pwm.setup(pin,1000,dc)                   -- set pin 3 as PWM pin and
7   pwm.start(pin)                           -- start PWM output
8   tmr.alarm(0,100,tmr.ALARM_AUTO,function()  -- repeat every 0.1 second
9     print("dc=",dc)                        -- debug dc
10    pwm.setduty(pin,dc)                    -- change PWM duty cycle (dc)
11    if ((dc < 19) and (ascdec == -1)) then
12      ascdec=1
13    end
14    if ((dc > 1003) and (ascdec == 1)) then
15      ascdec=-1
16    end
17    dc=(dc + (ascdec * inc))               -- increase or decrease dc by +/- 20
18  end)
```

圖 2-22(a)

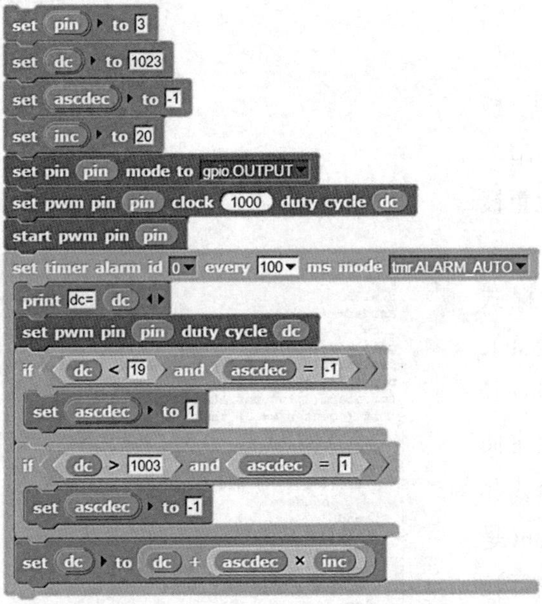

圖 2-22(b)

27

(2) 麵包板上接線方式如同步驟 a 的接法，接線結果如圖 2-23。

(3) 到 ESPlorer IDE 平台點擊左下方"Save to ESP"按鈕，將程式存檔並寫入到 NodeMCU 後執行，如圖 2-24。

圖 2-23

```
NodeMCU 1.5.1 build unspecified powered by Lua 5.1.4 on SDK 1.5.1(e(
lua: cannot open init.lua
> file.remove("script2.lua");
> file.open("script2.lua","w+");
> w = file.writeline
> w([[--set PWM pin to gpio pin 3]]);
> w([[pin = 3]]);
> w([[--set pin 3 to be OUTPUT pin]]);
> w([[gpio.mode(pin,gpio.OUTPUT)]]);
> w([[--set the highest brightness]]);
> w([[width = 1023]]);
> w([[--set the clock to 1000Hz]]);
> w([[clock = 1000]]);
> w([[-Set LED dimming step per timer]]);
> w([[dim = 50]]);
> w([[--set up pin 3 with 1000Hz]]);
> w([[pwm.setup(pin,clock,width)]]);
> w([[--start PWM output]]);
> w([[pwm.start(pin)]]);
> w([[--change the pulse width each 0.1 sec]]);
> w([[tmr.alarm(0,1000,1,function()]]);
> w([[    width = width- dim]]);
> w([[    -- LED darkest or LED brightest , change dimming step ]]
```

圖 2-24

c. 讓 LED 燈閃爍顯示(即一秒亮一秒暗)從最亮到最暗，再從最暗到最亮，如此重複不斷

(1) 修改步驟 a 的程式碼，利用增加兩個 state 和一個 count 設定，其中 state1 用來設定使 duty cycle 逐漸減少和逐漸遞增，而 count 是用來設定當美執行三次 duty cycle 的時候，就執行一次 state2，state2 則用來設定在每

```
1  counter=0              -- counter=0 PWM ascending or descending counter=1 LED off
2  pin=3                  -- set PWM to GPIO pin 3
3  dc=1023                -- set duty cycle to 1023
4  ascdec=-1              -- set dc change assending or decending
5  inc=20                 -- change delta value
6  gpio.mode(pin,gpio.OUTPUT)  -- set pin 3 to be an OUTPUT pin
7  pwm.setup(pin,1000,dc) -- set pin 3 as PWM pin and
8  pwm.start(pin)         -- start PWM output
9  tmr.alarm(0,100,tmr.ALARM_AUTO,function()  -- repeat every 0.1 second
10     if (counter == 0) then
11        counter=1              -- next cycle change to LED off
12        print("dc=",dc)        -- debug dc
13        pwm.setduty(pin,dc)    -- change PWM duty cycle (dc)
14        if ((dc < 19) and (ascdec == -1)) then
15           ascdec=1
16        end
17        if ((dc > 1003) and (ascdec == 1)) then
18           ascdec=-1
19        end
20        dc=(dc + (ascdec * inc))-- increase or decrease dc by +/- 20
21     else
22        counter=0              -- next cycle change to Ascending/Descending
23        print("dc=",0)
24        pwm.setduty(pin,0)
25     end
```

圖 2-25(a)

28

次 duty cycle 逐漸減少和逐漸遞增時同時開啟 LED 燈和關閉 LED 燈，如圖 2-25(a)。也可以使用 Snap4 Node MCU 讓 LED 燈閃爍顯示(即一秒亮一秒暗，從最亮到最暗，再從最暗到最亮，如此重覆不斷，如圖 2-25(b)。)

(2) 麵包板上接線方式如同步驟 a 的接法，接線結果如圖 2-26。

圖 2-25(b)

圖 2-26

(3) 到 ESPlorer IDE 平台點擊左下方 "Save to ESP" 按鈕，將程式存檔並寫入到 NodeMCU 後執行，如圖 2-27。

```
NodeMCU 1.5.1 build unspecified powered by Lua 5.1.4 on SDK 1.5.1(e67a03b)
lua: cannot open init.lua
> file.remove("led dimming_3.lua");
> file.open("led dimming_3.lua","w+");
> w = file.writeline
> w([[pin =3]]);
> w([[state1 ="down"      --state1 is used to change the value of duty cycle]]);
> w([[state2 ="on"        --state2 is used to set LED blinking ]]);
> w([[count=0]]);
> w([[gpio.mode(pin, gpio.OUTPUT]]);
> w([[width=1023]]);
> w([[clock=1000]]);
> w([[pwm.setup(pin,clock,width)]]);
> w([[pwm.start(pin)]]);
> w([[tmr.alarm(0,100,1,function()]]);
> w([[    if count ==2 then]]);
> w([[       count=0 ]]);
> w([[       if state2 == "on" then]]);
> w([[          --make pin output high]]);
> w([[          pwm.setduty(pin,width)]]);
> w([[          print("GPIO 3 light on")]]);
> w([[          state2 = "off"]]);
> w([[          -- if now is "open" close the Led and change state]]);
> w([[       else]]);
> w([[          --make pin output high]]);
> w([[          pwm.setduty(pin,0)]]);
> w([[          print("GPIO 3 light off")]]);
> w([[          state2 = "on"]]);
> w([[       end]]);
> w([[    else]]);
```

圖 2-27

29

4. 實驗成果：

a. 讓 LED 燈顯示從最亮到最暗，如此重複不斷。

(1) 因爲將 duty cycle 每次遞減 20，所以 ESPlorer 終端機會顯示 width 每次都遞減 20 的值，並且當遞減到 3 時直接重設 width 爲 1023，如圖 2-28：

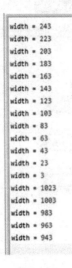

圖 2-28

圖 2-29

(2) 實際麵包板 LED 明暗情形。因此麵包板上的 LED 燈會由最亮逐漸變暗，當 LED 亮度最暗時，LED 又會變到最亮的狀態，如圖 2-29。

b. 讓 LED 燈顯示從最亮到最暗，再從最暗到最亮，如此重複不斷。

(1) 因爲將 duty cycle 每次遞減 50，所以 ESPlorer 終端機會顯示 width 每次都遞減 50 的值，但是當遞減到 23 時不是直接重設 width 爲 1023，而是逐漸再遞增 50，來達到漸亮和漸暗的功能，如圖 2-30。

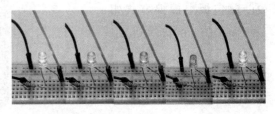

圖 2-30

(2) 實際麵包板 LED 明暗情形。麵包板
上的 LED 燈會由最亮逐漸變暗，當
LED 亮度最暗時，LED 又會逐漸變
亮到最亮的狀態，如圖 2-31。

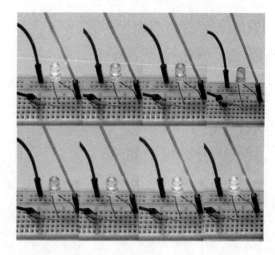

圖 2-31

c. 讓 LED 燈閃爍顯示(即一秒亮一秒暗)從最亮到最暗，再從最暗到最亮，如此重複不斷。

(1) 因為將 duty cycle 每次遞減 20，所以 ESPlorer 終端機會顯示 width 每次都遞減 20 的值，
但是當遞減到 3 時再逐漸遞增 20，並且每設定三次 duty cycle 值時關閉或開啟 LED 燈，
來達到漸亮漸暗以及同時明滅的功能，如圖 2-32：

```
width=843            width=1023
width=863            width=1023
width=883            width=1003
GPIO 3 light on      GPIO 3 light off
width=903            width=983
width=923            width=963
width=943            width=943
GPIO 3 light off     GPIO 3 light on
width=963            width=923
width=983            width=903
width=1003           width=883
GPIO 3 light on      GPIO 3 light off
width=1023           width=863
width=1023           width=843
width=1003           width=823
```

圖 2-32

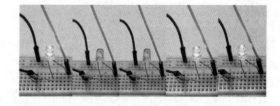

圖 2-33

(2) 實際麵包板 LED 明暗情形。麵包板上的 LED 燈會由最亮逐漸變暗，當 LED 亮度最暗時，
LED 又會逐漸變亮到最亮的狀態，並且同時會有明滅的情形，如圖 2-33。

主題A：使用 Fritzing IDE 平台做光感測器電路設計

1. 題目：使用 Fritzing 工具畫出 NodeMCU 與光感測器連接電路。
2. 實驗目標：使用光感測器並加入電阻範例練習 Fritzing 的電路設計，實際產生麵包板電路圖、電路概要圖以及 PCB 印刷板電路圖。
3. 實驗步驟：此實驗將分三個階段進行，

　　a. 產生麵包板電路圖。

　　b. 產生電路概要圖。

　　c. 產生 PCB 印刷板電路圖。

a. 產生麵包板電路圖

　　(1) 開啟 Fritzing 工具，點選"麵包板"，如圖 3-1。

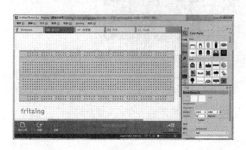
圖 3-1

圖 3-2

　　(2) 在右方搜尋窗格輸入"NodeMCU"，找到 NodeMCU 元件。將 NodeMCU 往麵包板上拖曳並放置在麵包板上並完成接線，如圖 3-2。此處為了之後元件接線美觀，將 GND 接到麵包板上方的藍邊，3V3 接到麵包板下方的紅邊。

　　(3) 如圖 3-3 搜尋"LDR"，將光感測器拖曳至麵包板上，如圖 3-4。

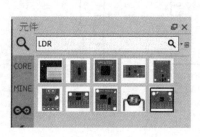
圖 3-3

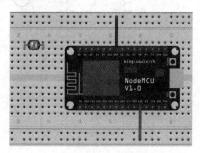
圖 3-4

(4) 將光感測器的一邊接腳接到 NodeMCU 上的 A0 數位接孔，一邊接腳接到麵包板上的 GND，
如圖 3-5。

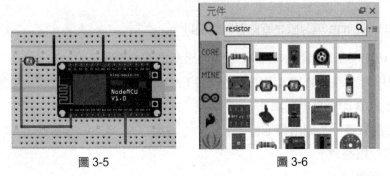

圖 3-5   圖 3-6

(5) 如圖 3-6 搜尋"resistor"，將電阻拖曳至麵包板上，電阻的一邊要和光感測器的一邊接孔在
同一行上，如圖 3-7。

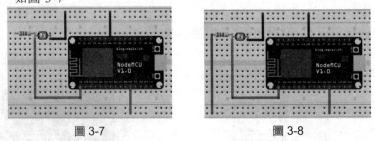

圖 3-7   圖 3-8

(6) 將電阻的另一端接腳接到麵包板上的電源接孔，如圖 3-8。

b. 產生電路概要圖

點選上方的"概要圖"標籤即可看到自動幫我們建好的電路概要圖，如圖 3-9 左。將元件以及線路
擺放至較美觀的位置，並將線路佈線完成，如圖 3-9 右圖。

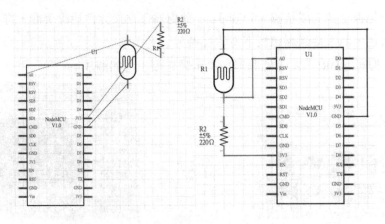

圖 3-9

c. 產生 PCB 印刷板電路圖

　點選上方的"PCB"標籤看到自動幫我們建好的 PCB 印刷板電路圖，接著將元件以及線路擺放至較美觀的位置，並將線路佈線完成，如圖 3-10。

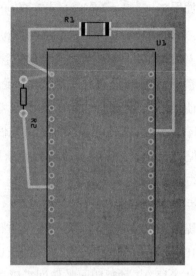

圖 3-10

4. 實驗成果：

　(1) 麵包板電路如圖 3-11。

　(2) 電路概要圖如圖 3-12。

　(3) PCB 印刷板電路圖如圖 3-13。

圖 3-11

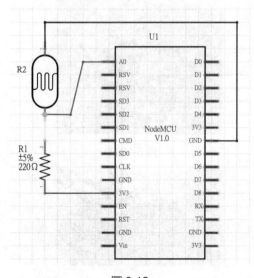

圖 3-12

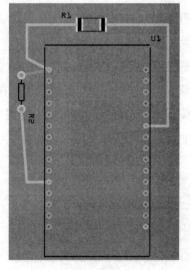

圖 3-13

35

主題 B：使用 Snap4NodeMCU 與 ESPlorer IDE 平台做光感測器程式設計

1. 題目：實際利用麵包板將 NodeMCU 以及 LED 進行線路連接，並使用 ESPlorer IDE 平台撰寫讀取光感測器的值，並觀察以及實驗下述兩點：

    (1) 當手不蓋/半蓋/蓋住光感測器，觀察終端機顯示的變化。

    (2) 使用手機"手電筒"App 照射感測器，觀察終端機顯示的變化。

2. 實驗目標：使用讀取光感測器的值範例程式練習 Lua 的程式設計，並實際觀察麵包板接線完成後，程式進行時 ESPlorer 終端機顯示畫面，讓使用者實際了解當光源變化時，光感測器實際讀取值的變化情形。

3. 實驗步驟：

    (1) 點擊 ESPlorer.bat 工具，進入開發環境 IDE。在左方的程式撰寫區撰寫 Lua 程式，每 0.1 秒呼叫函數傳回光感測器的讀取值，如圖 3-14(a)。也可以使用 Snap4NodeMCU 讀取光感測器，如圖 3-14(b)。

```
1  tmr.alarm(0,100,tmr.ALARM_AUTO,function()
2      aValue=adc.read(0)          -- read analog value from pin 0
3      print("lightness=",aValue)  -- debug analog value
4  end)
```
```
1  pin = 0
2  tmr.alarm(0,200,1,function()
3      print("adc = "..adc.read(pin))
4  end)
```

圖 3-14(a)

圖 3-14(b)

    (2) 將 NodeMCU 插入到麵包板上，再利用杜邦線把 NodeMCU 的電源(3V3)接腳和接地(GND)接腳接到麵包板上的電源和接地接孔，如圖 3-15。

    (3) 將光感測器模組(圖 3-16)的 A0 接腳接到 NodeMCU 的 A0 接腳，VCC 和 GND 接腳分別接到 NodeMCU 上的電源和接地接孔，如圖 3-17。

圖 3-15

圖 3-16

圖 3-17

(4) 到 ESPlorer IDE 平台點擊左下方"Save to ESP"按鈕，將程式存檔並寫入到 NodeMCU 後執行。

4. 實驗成果：

(1) 當手不蓋住光感測器，終端機顯示的變化如圖 3-16：

(2) 當手半蓋住光感測器，終端機顯示的變化如圖 3-19：

```
> dofile("Lab3_1.lua");
> adc = 145
adc = 146
adc = 146
adc = 146
adc = 144
adc = 147
adc = 147
adc = 148
adc = 148
adc = 145
adc = 149
adc = 149
adc = 149
adc = 149
```

圖 3-18

```
adc = 391
adc = 390
adc = 393
adc = 393
adc = 396
adc = 392
adc = 394
adc = 394
adc = 391
adc = 395
adc = 392
adc = 392
adc = 391
adc = 393
adc = 391
adc = 385
adc = 372
```

圖 3-19

(3) 當手蓋住光感測器，終端機顯示的變化如圖 3-20：

(4) 使用手機"手電筒"App 照射感測器，終端機顯示的變化如圖 3-21：

經由上述實驗可以實際瞭解，透過分壓原理利用電阻值來得知亮度，當亮度越亮讀到的值越小，當亮度越暗時讀到的值越大。

```
adc = 64
adc = 64
adc = 65
adc = 64
adc = 64
adc = 64
adc = 64
adc = 64
adc = 64
adc = 64
adc = 64
adc = 64
adc = 64
adc = 63
adc = 63
adc = 63
adc = 63
adc = 64
adc = 63
adc = 63
adc = 63
```

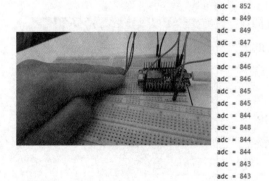

```
adc = 852
adc = 849
adc = 849
adc = 847
adc = 847
adc = 846
adc = 846
adc = 845
adc = 845
adc = 848
adc = 844
adc = 844
adc = 843
adc = 843
```

圖 3-20

圖 3-21

主題 C：使用 Snap4NodeMCU 與 ESPlorer IDE 平台做光感測器結合智慧燈光程式設計

1. 題目：使用 ESPlorer IDE 平台撰寫讀取光感測器的值，根據環境光源的改變，調整 LED 燈的明亮。

2. 實驗目標：以 LED 燈模擬可供生活照明的燈光，當環境光源變暗時，需要提高照明的亮度，以保持光線充足；當環境光源變亮時，則關閉 LED 燈的照明。

3. 實驗步驟：此實驗將分兩個階段進行，

   a. 當光度不足時，自動啟動 LED 燈。

   b. 當光度足夠時關閉 LED 燈；光度不足時調整 LED 燈亮度。

a. 當光度不足時，自動啟動 LED 燈

   (1) 點擊 ESPlorer.bat 工具，進入開發環境 IDE。在左方的程式撰寫區撰寫 Lua 程式，設定變數 ambient 為環境光源的亮度全暗時的值，當光感測器的光度比環境光源的值大時亮燈，反之則暗燈，如圖 3-22(a)。也可以使用 Snap4NodeMCU 實現智慧燈光設計，如圖 3-22(b)。

```
1  led=4                                -- set led to pin 4
2  gpio.mode(led,gpio.OUTPUT)           -- set pin 4 as output pin
3  ambient=500                          -- set a threshold value
4  tmr.alarm(0,100,tmr.ALARM_AUTO,function() -- repeat every 0.1 second
5      bright=adc.read(0)               -- read analog value from pin 0
6      print("lightness=",bright)       -- debug analog value
7      if (bright > ambient) then       -- if the current lightness is not good enought
8          gpio.write(led,gpio.HIGH)    -- trun on LED (pin 4)
9      else
10         gpio.write(led,gpio.LOW)     -- trun off LED (pin 4)
11     end
12  end)
```

```
1   pin = 0
2   -- set LED to D1
3   led = 1
4   -- set the brightness of environment
5   ambient = 840
6   tmr.alarm(0,500,1,function()
7       bright = adc.read(pin)
8       print("bright = " ..bright)
9       --if brightness is bigger than ambient, than open the Light
10      if bright >= ambient then
11          gpio.write(led, gpio.HIGH)
12      --otherwise, close the light
13      else
14          gpio.write(led,gpio.LOW)
15      end
16  end)
17
```

圖 3-22(a)

圖 3-22(b)

   (2) 接續主題 B 的實際麵包板接線，再將 LED 燈插入到麵包板上，將長邊接腳接到 NodeMCU 的 D1 接腳，短邊接腳和電阻相接，電阻的另一端接腳則接到麵包板上的接地接孔，如圖 3-23。

   (3) 到 ESPlorer IDE 平台點擊左下方"Save to ESP"按鈕，將程式存檔並寫入到 NodeMCU 後執行。

圖 3-23

b. 當光度足夠時關閉 LED 燈；光度不足時調整 LED 燈亮度

  (1) 修改步驟 a 的程式碼，設定變數 ambient 為環境光源的亮度足夠時的值。當光感測器的光度比環境光源 ambient 的值大時會隨著光度的值改變 LED 的亮度；而當光感測器的光度比環境光源 ambient 的值小時，會將 LED 燈熄滅。程式碼範例如圖 3-24。

  (2) 麵包板上接線方式如同步驟 a 的接法，接線結果如圖 3-25。

  (3) 到 ESPlorer IDE 平台點擊左下方"Save to ESP"按鈕，將程式存檔並寫入到 NodeMCU 後執行。

4. 實驗成果：

  a. 當光度不足時，自動啟動 LED 燈

    (1) 一開始執行程式時，終端機畫面如圖 3-26。

```
1   pin = 0
2   -- set LED to D1
3   led = 1
4   -- set the brightness of environment
5   ambient = 150
6   -- set the highest brightness
7   width = 0
8   --set the clock to 1000Hz
9   clock = 1000
10  pwm.setup(led,clock,width)
11  --start PWM output
12  pwm.start(led)
13  tmr.alarm(0,500,1,function()
14      bright = adc.read(pin)
15      --if brightness is bigger than ambient
16      if bright <= ambient then
17          width = 0
18      else
19          width = bright
20      end
21
22      pwm.setduty(led,width)
23      print("bright = " ..bright ..",width = " ..width)
24  end)
```

圖 3-24

圖 3-25

```
bright = 125
bright = 123
bright = 121
bright = 121
bright = 121
bright = 122
bright = 123
bright = 124
bright = 126
bright = 128
bright = 131
bright = 132
bright = 131
bright = 126
bright = 124
bright = 122
bright = 120
bright = 120
bright = 120
bright = 121
bright = 123
```

圖 3-26

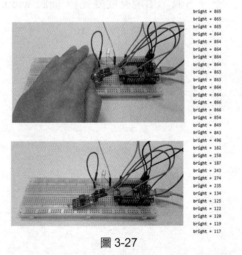

```
bright = 865
bright = 865
bright = 865
bright = 864
bright = 864
bright = 864
bright = 864
bright = 863
bright = 863
bright = 864
bright = 864
bright = 866
bright = 866
bright = 854
bright = 849
bright = 843
bright = 496
bright = 162
bright = 158
bright = 187
bright = 243
bright = 274
bright = 235
bright = 134
bright = 125
bright = 122
bright = 120
bright = 119
bright = 117
```

圖 3-27

    (2) 當用手將光感測器遮住時，則會偵測到光度不夠，因此就將 LED 燈打開，反之就將 LED 燈關閉。實際操作圖如圖 3-27，虛線上方為將手把光感測器度遮住模擬光線不足時，會將 LED 燈打開；虛線下方則是當光線足夠時，就會把 LED 燈關閉。

b. 當光度足夠時關閉 LED 燈；光度不足時調整 LED 燈亮度

(1) 一開始執行程式時，終端機畫面如圖 3-28。

```
> dofile("Lab3_3-4.lua");
> bright = 132,width = 0
bright = 135,width = 0
bright = 137,width = 0
bright = 140,width = 0
bright = 141,width = 0
bright = 141,width = 0
bright = 140,width = 0
bright = 137,width = 0
bright = 133,width = 0
bright = 131,width = 0
bright = 135,width = 0
```

```
bright = 137,width = 0
bright = 133,width = 0
bright = 131,width = 0
bright = 135,width = 0
bright = 293,width = 293
bright = 368,width = 368
bright = 575,width = 575
bright = 692,width = 692
bright = 817,width = 817
bright = 900,width = 900
bright = 920,width = 920
bright = 934,width = 934
bright = 933,width = 933
bright = 934,width = 934
bright = 934,width = 934
bright = 933,width = 933
bright = 932,width = 932
bright = 907,width = 907
bright = 691,width = 691
bright = 533,width = 533
bright = 412,width = 412
bright = 313,width = 313
bright = 270,width = 270
bright = 181,width = 181
bright = 165,width = 165
bright = 130,width = 0
bright = 132,width = 0
```

圖 3-28   圖 3-29

(2) 如圖 3-29，虛線上方為將手把光感測器度逐漸遮住時，隨著 bright 逐漸變大(表示環境亮度變暗)，而將 width 值逐漸變大亦即 LED 亮度逐漸變亮；虛線下方則是 bright 逐漸變小(表示環境亮度變亮)，而將 width 值逐漸變小直到亮度夠亮，就會把 LED 燈關閉。

主題 A：使用 Fritzing IDE 平台做 PIR 感測器電路設計

1. 題目：使用 Fritzing 工具畫出 NodeMCU 與 PIR 感測器連接電路。

2. 實驗目標：使用 PIR 感測器範例練習 Fritzing 的電路設計，實際產生麵包板電路圖、電路概要圖以及 PCB 印刷板電路圖。

3. 實驗步驟：此實驗將分三個階段進行，

   a. 產生麵包板電路圖。

   b. 產生電路概要圖。

   c. 產生 PCB 印刷板電路圖。

a. 產生麵包板電路圖

   (1) 開啟 Fritzing 工具，點選"麵包板"，如圖 4-1。

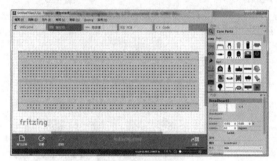 

圖 4-1   圖 4-2

   (2) 在右方搜尋窗格輸入"NodeMCU"，將 NodeMCU 拖曳至麵包板上並完成接線，如圖 4-2。

   (3) 由於 Fritzing 沒有 PIR 這個元件，所以可以經由網頁下載相關元件，網址：
       https://github.com/adafruit/Fritzing-Library，如圖 4-3。

圖 4-3

(4) 在右方空白處按右鍵，選擇"Import"，如圖 4-4。

圖 4-4

圖 4-5

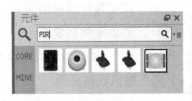
圖 4-6

(5) 選擇在上一步中所下載的檔案位置下的 parts 子目錄中的 PIR sensor.fzpz，如圖 4-5。

(6) 之後可以藉由搜尋元件 PIR 即可，如圖 4-6。

(7) 將 PIR 拖曳至麵包板上，如圖 4-7。

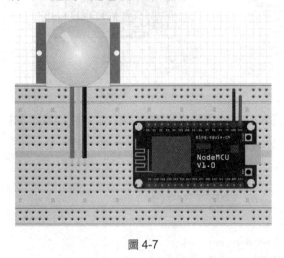
圖 4-7

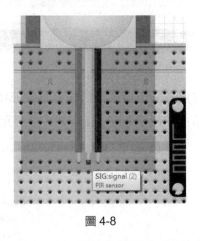
圖 4-8

(8) 移動到 PIR 在麵包板上的接腳，會解釋該接腳的功能為何可以知道該隻接腳的功用，如圖 4-8 黃色接腳是 signal 接腳。

(9) 將 PIR 的黃色接腳接到 NodeMCU 上的 D4 數位接孔,黑色接腳接到麵包板上的 GND,紅色接腳接到 NodeMCU 上的 Vin 接孔,如圖 4-9。

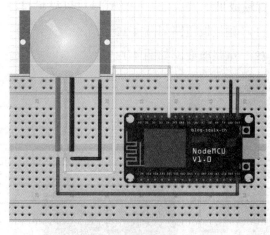

圖 4-9

b. 產生電路概要圖

點選上方的"概要圖"標籤即可看到自動幫我們建好的電路概要圖,將元件以及線路擺放至較美觀的位置,並將線路佈線完成,如圖 4-10。

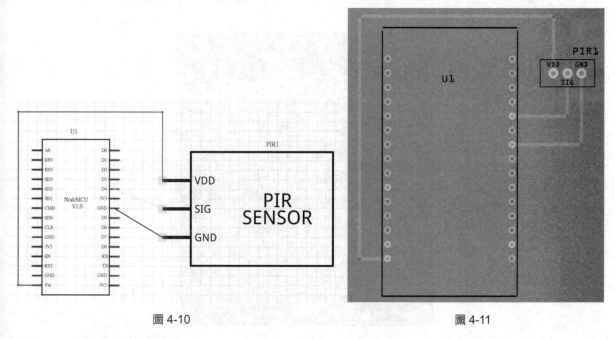

圖 4-10

圖 4-11

c. 產生 PCB 印刷板電路圖

點選上方的"PCB"標籤看到自動幫我們建好的 PCB 印刷板電路圖,將元件以及線路擺放至較美觀的位置,並將線路佈線完成,如圖 4-11。

4. 實驗成果：

    (1) 麵包板電路如圖 4-12。

    (2) 電路概要圖如圖 4-13。

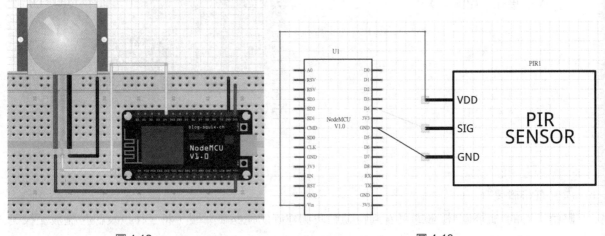

圖 4-12   圖 4-13

    (3) PCB 印刷板電路圖如圖 4-14。

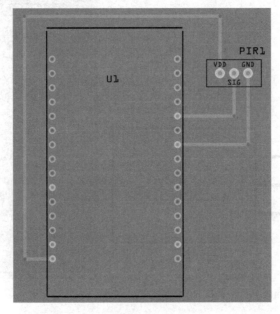

圖 4-14

主題 B：使用 Snap4NodeMCU 與 ESPlorer IDE 平台做 PIR 感測器程式設計

1. 題目：實際利用麵包板將 NodeMCU 以及 PIR 進行線路連接，並使用 ESPlorer IDE 平台撰寫讀取 PIR 感測器的值，並觀察以及實驗下述三點：

    (1) 當揮手/蓋住人體感應模組，觀察終端機顯示的變化。

    (2) 調整延遲時間 (Delay time)，觀察終端機顯示的變化。

    (3) 調整感應距離 (Sensing distance)，觀察終端機顯示的變化。

2. 實驗目標：使用讀取 PIR 感測器的值之範例程式練習 Lua 的程式設計，並實際觀察麵包板接線完成後，程式進行時 ESPlorer 終端機顯示畫面，讓使用者實際了解當 PIR 偵測到變化時，實際讀取值的變化情形。

3. 實驗步驟：

    (1) 點擊 ESPlorer.bat 工具，進入開發環境 IDE。在程式撰寫區撰寫 Lua 程式，每 0.5 秒呼叫函數傳回 PIR 感測器的讀取值，如圖 4-15(a)。也可以使用 Snap4NodeMCU 讀取 PIR 感測器，如圖 4-15(b)。

```
1  PIRPin=5                               -- set PIRPin to pin 5
2  gpio.mode(PIRPin,gpio.INPUT)           -- set pin 5 as digital input pin
3  i=1                                    -- set variable i to 1
4  tmr.alarm(0,500,tmr.ALARM_AUTO,function()  -- repeat every 0.5 second
5    state=gpio.read(PIRPin)              -- read PIR value and set to variable state
6    print(i," PIR=",state)               -- debug state
7    i=(i + 1)                            -- increase i by 1
8  end)
```

圖 4-15(a)

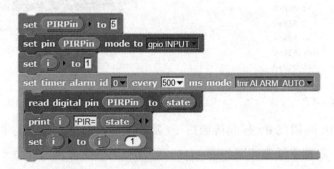

圖 4-15(b)

(2) 將 NodeMCU 插入到麵包板上，再利用杜邦線把 NodeMCU 的電源(3V3)接腳和接地(GND)接腳接到麵包板上的電源和接地接孔，並將 PIR 感測器的 OUT 接腳接到 NodeMCU 上的 D5 數位接孔，GND 接腳接到麵包板上的 GND，VCC 接腳接到 NodeMCU 上的 Vin 接孔，如圖 4-16。

(3) 到 ESPlorer IDE 平台點擊左下方"Save to ESP" 按鈕，將程式存檔並寫入到 NodeMCU 後執行。

圖 4-16

4. 實驗成果：

(1) 當揮手/蓋住人體感應模組，終端機顯示的變化如圖 4-17，當 PIR 附近物體移動時其輸出為 1，離開後才會將輸出改為 0。

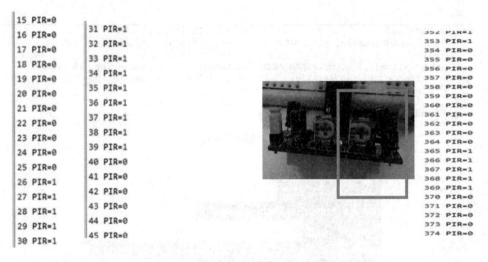

```
15 PIR=0      31 PIR=1                            352 PIR=1
16 PIR=0      32 PIR=1                            353 PIR=1
17 PIR=0      33 PIR=1                            354 PIR=0
18 PIR=0      34 PIR=1                            355 PIR=0
19 PIR=0      35 PIR=1                            356 PIR=0
20 PIR=0      36 PIR=1                            357 PIR=0
21 PIR=0      37 PIR=1                            358 PIR=0
22 PIR=0      38 PIR=1                            359 PIR=0
23 PIR=0      39 PIR=1                            360 PIR=0
24 PIR=0      40 PIR=0                            361 PIR=0
25 PIR=0      41 PIR=0                            362 PIR=0
26 PIR=1      42 PIR=0                            363 PIR=0
27 PIR=1      43 PIR=0                            364 PIR=0
28 PIR=1      44 PIR=0                            365 PIR=1
29 PIR=1      45 PIR=0                            366 PIR=1
30 PIR=1                                          367 PIR=1
                                                  368 PIR=1
                                                  369 PIR=1
                                                  370 PIR=0
                                                  371 PIR=0
                                                  372 PIR=0
                                                  373 PIR=0
                                                  374 PIR=0
```

圖 4-17   圖 4-18

(2) 一開始蓋住 PIR 時值為 0，移開後值為 1。當一開始沒有調整延遲時間時，終端機顯示的變化如圖 4-18。當調整延遲時間如圖 4-19，可以發現移開後再蓋住時，PIR 值為 1 的延遲時間比較長。

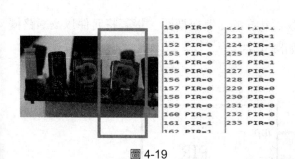

圖 4-19 　　　　　　　　　　　　　　　　圖 4-20

(3) 將左邊旋鈕順時針旋轉調整感應距離(Sensing distance)如圖 4-20，當距離遠一點時 PIR 值也會改變為 1。

主題 C：使用 Fritzing、Snap4NodeMCU 與 ESPlorer IDE 平台做 PIR 感測器結合智慧燈光之軟硬體設計

1. 題目：使用 ESPlorer IDE 平台撰寫讀取 PIR 感測器的值，根據需求自動開啟 LED 燈。

2. 實驗目標：將 PIR 結合 LED 燈模擬智慧燈光的實際情形。

3. 實驗步驟：此實驗將分三個階段進行，

　a. 使用 Fritzing 實際產生麵包板電路圖、電路概要圖以及 PCB 印刷板電路圖，以及實際的麵包板接線圖。

　b. 當感應人體時，自動啟動 LED 燈。

　c. 當感應人體時且光線不足時，自動啟動 LED 燈。

a. 使用 Fritzing 實際產生麵包板電路圖、電路概要圖以及 PCB 印刷板電路圖，以及實際的麵包板接線圖

(1) 延續主題 B，將 LED 燈和電阻拉到麵包板上，如圖 4-21。

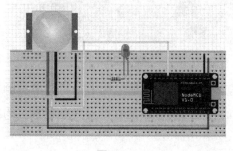

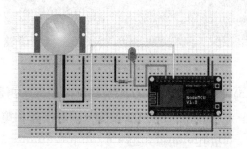

圖 4-21 　　　　　　　　　　　　　　　　圖 4-22

(2) 將 LED 的長邊接腳接到 NodeMCU 上的 D2 數位接孔，電阻的另一端接腳接到麵包板上的 GND，如圖 4-22。

(3) 點選上方的"概要圖"標籤即可看到自動幫我們建好的電路概要圖，接著將元件以及線路擺放至較美觀的位置，並將線路佈線完成，如圖 4-23。

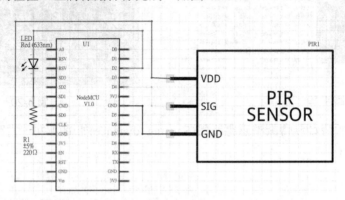

圖 4-23

(4) 點選上方的"PCB"標籤看到自動幫我們建好的 PCB 印刷板電路圖，再將元件以及線路擺放至較美觀的位置，並將線路佈線完成，如圖 4-24。

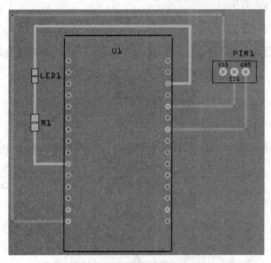

圖 4-24

圖 4-25

(5) 接續主題 B 的實際麵包板接線，再將 LED 燈和電阻插入到麵包板上，將長邊接腳接到 NodeMCU 的 D2 接腳，短邊接腳和電阻相接，而電阻的另一端接腳則接到麵包板上的接地接孔，如圖 4-25。

b. 當感應人體時，自動啟動 LED 燈

(1) 點擊 ESPlorer.bat 工具，進入開發環境 IDE。在左方的程式撰寫區撰寫 Lua 程式，設定 LED 燈接腳為 2，當 PIR 感測器的值為 1，表示有人移動時亮燈，反之則暗燈，如圖 4-26(a)。也可以使用 Snap4NodeMCU 當感應人體時，自動啟動 LED 燈，如圖 4-26(b)。

```lua
1   led=2                                  -- set led to pin 2
2   gpio.mode(led,gpio.OUTPUT)             -- set pin 2 as digital output pin
3   PIRPin=5                               -- set PIRPin to pin 5
4   gpio.mode(PIRPin,gpio.INPUT)           -- set pin 5 as digital input pin
5   i=1                                    -- set variable i to 1
6   tmr.alarm(0,500,tmr.ALARM_AUTO,function()   -- repeat every 0.5 second
7     state=gpio.read(PIRPin)              -- read PIR value and set to variable state
8     print(i," PIR=",state)              -- debug state
9     if (state == 1) then                 -- if someone is detected
10       gpio.write(led,gpio.HIGH)         -- turn on LED
11    else
12       gpio.write(led,gpio.LOW)          -- turn off LED
13    end
14    i=(i + 1)                            -- increase i by 1
15  end)
```

圖 4-26(a)

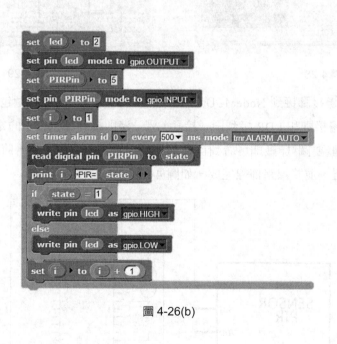

圖 4-26(b)

(2) 實際麵包板接線方式如同步驟 a，詳細接線圖如圖 4-27。

(3) 到 ESPlorer IDE 平台點擊左下方"Save to ESP"按鈕，將程式存檔並寫入到 NodeMCU 後執行。

圖 4-27

c. 當感應人體時且光線不足時，自動啟動 LED 燈

(1) 延續步驟 a 的麵包板電路圖，將 LDR(光感測器)以及 resistor(電阻)拉到麵包板上，如圖 4-28。

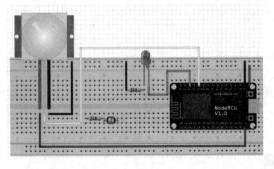 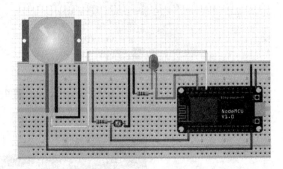

圖 4-28   圖 4-29

(2) 將 LDR 的一邊接腳接到 NodeMCU 的 A0 接腳，另一邊接腳接到麵包板上的接地接孔，將 resistor 的一邊接腳和 LDR 接到同一接孔，另一邊則接到麵包板上的電源接孔，如圖 4-29。

(3) 點選上方的"概要圖"標籤即可看到自動幫我們建好的電路概要圖，將元件以及線路擺放至較美觀的位置，並將線路佈線完成，如圖 4-30。

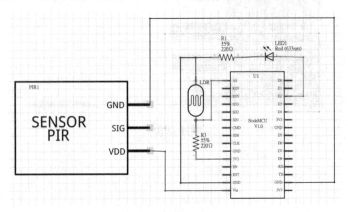

圖 4-30

(4) 點選上方的"PCB"標籤看到自動幫我們建好的 PCB 印刷板電路圖，將元件以及線路擺放至
較美觀的位置，並將線路佈線完成，如圖 4-31。

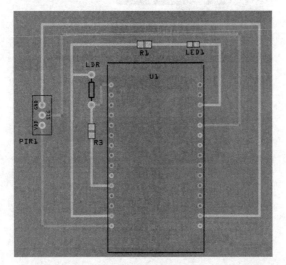

圖 4-31

(5) 修改步驟 a 的程式碼，設定變數 ambient 為環境光源的亮度足夠時的值，lightpin 表示光感
測器的接腳。當偵測到有人而且光感測器的光度比環境光源 ambient 的值大時，表示光度
不夠量因此會將 LED 燈打開；而當光感測器的光度比環境光源 ambient 的值小時，則會將
LED 燈熄滅。程式碼範例如圖 4-32(a)。也可以使用 Snap4NodeMCU 加以實現，如圖 4-32(b)。

```
1   ambient=800                        -- set ambient to pin 800
2   led=2                              -- set led to pin 2
3   gpio.mode(led,gpio.OUTPUT)         -- set pin 2 as digital output pin
4   PIRpin=5                           -- set PIRPin to pin 5
5   gpio.mode(PIRpin,gpio.INPUT)       -- set pin 5 as digital input pin
6   i=1                                -- set variable i to 1
7   tmr.alarm(0,500,tmr.ALARM_AUTO,function()
8     state=gpio.read(PIRpin)
9     bright=adc.read(0)
10    print(i," PIR=",state,"bright=",bright)
11    if ((state == 1) and (bright < ambient)) then
12      gpio.write(led,gpio.HIGH)
13    else
14      gpio.write(led,gpio.LOW)
15    end
16    i=(i + 1)
17    state=gpio.read(PIRpin)          -- read PIR value and set to variable state
18    bright=adc.read(0)               -- read brightness to bright variable
19    print(i," PIR=",state,"bright=",bright)  -- debug state and bright
20    if ((state == 1) and (bright < ambient)) then   -- if someone is detected and it is too dark
21      gpio.write(led,gpio.HIGH)      -- turn on LED
22    else
23      gpio.write(led,gpio.LOW)       -- turn off LED
24    end
25    i=(i + 1)                        -- increase i by 1
26  end)
```

圖 4-32(a)

51

```
set ambient to 800
set led to 4
set pin led mode to gpio.OUTPUT
set PIRPin to 5
set pin PIRPin mode to gpio.INPUT
set i to 1
set timer alarm id 0 every 500 ms mode tmr.ALARM_AUTO
  read digital pin PIRPin to state
  read analog pin to bright
  print i "PIR=" state bright= bright
  if state = 1 and bright < ambient
    write pin led as gpio.HIGH
  else
    write pin led as gpio.LOW
  set i to i + 1
```

圖 4-32(b)

(6) 將光感測器模組 A0 接腳接到 NodeMCU 的 A0 接腳，VCC 和 GND 接腳分別接到 NodeMCU
上的電源和接地接孔，如圖 4-33。

圖 4-33

(7) 到 ESPlorer IDE 平台點擊左下方"Save to ESP"按鈕，將程式存檔並寫入到 NodeMCU 後執
行。

52

4. 實驗成果：

a. 使用 Fritzing 實際產生麵包板電路圖、電路概要圖以及 PCB 印刷板電路圖，以及實際的麵包板接線圖

    (1) 麵包板電路圖如圖 4-34。

    (2) 電路概要圖如圖 4-35。

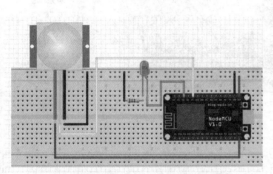

圖 4-34

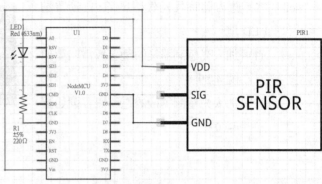

圖 4-35

    (3) PCB 印刷板電路圖如圖 4-36。

    (4) 實際的麵包板接線圖如圖 4-37。

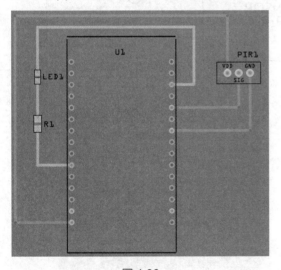

圖 4-36

圖 4-37

b. 當感應人體時，自動啟動 LED 燈

當紅外線偵測到沒有人時，PIR 值為 0 且 LED 燈不亮；當偵測到有人時，PIR 值為 1 且 LED 燈亮起，實際畫面如圖 4-38。

c. 當感應人體時且光線不足時，自動啟動 LED 燈

(1) 加入光感測器後的麵包板電路圖如圖 4-39。

(2) 加入光感測器後的電路概要圖如圖 4-40。

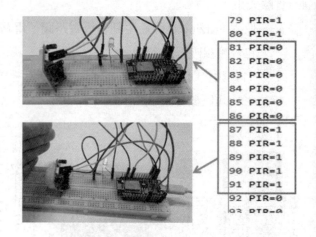

圖 4-38

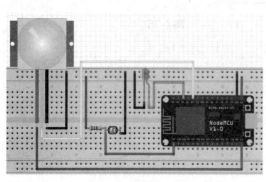

圖 4-39

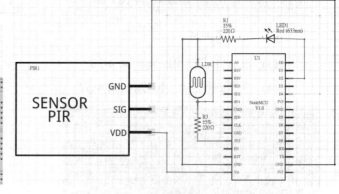

圖 4-40

(3) 加入光感測器後的 PCB 印刷板電路圖如圖 4-41。

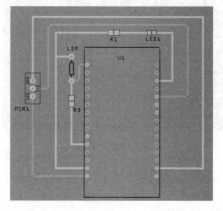

圖 4-41

(4) 加入光感測器後的實際的麵包板接線圖如圖 4-42。

圖 4-42

圖 4-43

(5) 當紅外線偵測沒有人而且光度充足時，實際畫面如圖 4-43。

(6) 當紅外線偵測沒有人而且光度不夠時，實際畫面如圖 4-44。

圖 4-44

圖 4-45

(7) 當紅外線偵測有人而且光度充足時，實際畫面如圖 4-45。

(8) 當紅外線偵測有人而且光度不夠時，實際畫面如圖 4-46。

圖 4-46

主題 A：使用 Fritzing IDE 平台做矩陣鍵盤感測器電路設計

1. 題目：使用 Fritzing 工具畫出 NodeMCU 與矩陣鍵盤感測器連接電路。
2. 實驗目標：使用矩陣鍵盤感測器範例練習 Fritzing 的電路設計，實際產生麵包板電路圖、電路概要圖以及 PCB 印刷板電路圖。
3. 實驗步驟：此實驗將分三個階段進行，
   a. 產生麵包板電路圖。
   b. 產生電路概要圖。
   c. 產生 PCB 印刷板電路圖。
a. 產生麵包板電路圖
   (1) 開啟 Fritzing 工具，點選"麵包板"，如圖 5-1。
   (2) 在右方搜尋窗格輸入"NodeMCU"，將 NodeMCU 拖曳至麵包板上並完成接線，如圖 5-2。
   (3) 由於 Fritzing 沒有 4×4 Keypad 這個元件，所以可以經由網頁下載相關元件，網址：https://github.com/ciromattia/Fritzing-Library，如圖 5-3。

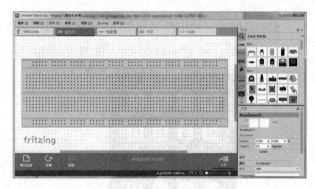

圖 5-1

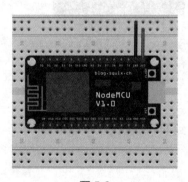

圖 5-2

圖 5-3

57

(4) 按右鍵" Import"，選擇所下載的元件存放位置，選擇"KEYPAD 4x4.fzpz"匯入元件，如圖 5-4。

圖 5-4   圖 5-5

(5) 之後可以藉由搜尋元件 Keypad 即可。將 Keypad 拖曳至麵包板上，如圖 5-5。

(6) 將 Keypad 的 col1 到 col4 接到 NodeMCU 上的 D1 到 D4 數位接孔，如圖 5-6。

圖 5-6   圖 5-7

(7) 將 Keypad 的 row1 到 row4 接到 NodeMCU 上的 D5 到 D8 數位接孔，如圖 5-7。

b. 產生電路概要圖

點選上方的"概要圖"標籤即可看到自動幫我們建好的電路概要圖，將元件以及線路擺放至較美觀的位置，並將線路佈線完成，如圖 5-8。

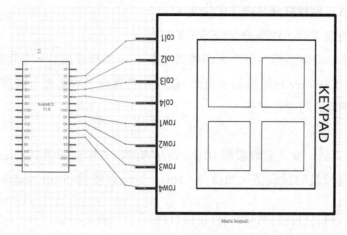

圖 5-8

c. 產生 PCB 印刷板電路圖

由於目前所提供的 4×4 Keypad 元件無法提供 PCB 接線，因此在本實驗中不提供 PCB 接線圖。

4. 實驗成果：

(1) 麵包板電路如圖 5-9。

(2) 電路概要圖如圖 5-10。

圖 5-9

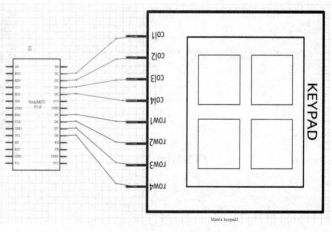

圖 5-10

主題 B：使用 Snap4NodeMCU 與 ESPlorer IDE 平台做鍵盤感測器程式設計

1. 題目：實際利用麵包板將 NodeMCU 以及鍵盤感測器進行線路連接，並使用 ESPlorer IDE 平台撰寫讀取鍵盤感測器的值，並觀察以及實驗下述兩點：

    (1) 按下鍵盤就放，觀察終端機顯示的變化。

    (2) 長時間按下鍵盤才放，觀察終端機顯示的變化。

2. 實驗目標：使用讀取鍵盤感測器值之範例程式練習 Lua 的程式設計，並實際觀察麵包板接線完成後，程式進行時 ESPlorer 終端機顯示畫面，讓使用者實際了解當按下鍵盤感測器上的按鍵時，鍵盤感測器實際讀取值的變化情形。

3. 實驗步驟：

(1) 點擊 ESPlorer.bat 工具，進入開發環境 IDE。在左方的程式撰寫區撰寫 Lua 程式，每 0.2 秒呼叫函數傳回按下鍵盤按鍵的讀取值，如圖 5-11(a)。也可以使用 Snap4NodeMCU 加以實現，如圖 5-11(b)。

```
1   -- define 4x4 keypad name
2   keys={{"1","2","3","A"},{"4","5","6","B"},{"7","8","9","C"},{"*","0","#","D"}}
3   for i=1,8 do
4      gpio.mode(i,gpio.OUTPUT)    -- set pin 1~4 as output pin
5      gpio.write(i,gpio.LOW)      -- set pin 1~8 to low
6   end
7   for i=5,8 do
8      gpio.mode(i,gpio.INPUT)     -- set pin 5~8 as input pin
9   end
10  tmr.alarm(0,200,tmr.ALARM_AUTO,function()   -- repeat very 0.2 second
11     -- key scan algorithm
12     for r=1,4 do                 -- loop for each row pin
13        gpio.write(r,gpio.HIGH)    -- set row pin to high
14        for c=1,4 do               -- loop for each column pin
15           dValue=gpio.read((c + 4)) -- read value of column pin
16           if (dValue == 1) then    -- if the value is 1 then a key is pressed
17              print("r=",r,"c=",c,"key=",keys[r][c]) -- print out which key
18           end
19        end
20        gpio.write(r,gpio.LOW)     -- set row pin to low
21     end
22  end)
```

圖 5-11(a)

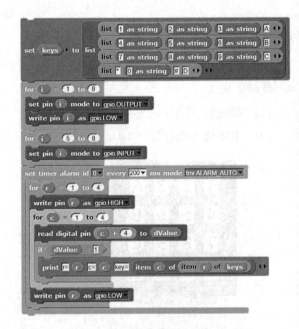

圖 5-11(b)

(2) 將 NodeMCU 插入到麵包板上，再利用杜邦線把 NodeMCU 的電源(3V3)接腳和接地(GND)接腳接到麵包板上的電源和接地接孔，再將鍵盤感測器模組的 col1 到 col4 接到 NodeMCU 上的 D1 到 D4 數位接孔，row1 到 row4 接到 NodeMCU 上的 D5 到 D8 數位接孔，如圖 5-12。

圖 5-12

(3) 到 ESPlorer IDE 平台點擊左下方"Save to ESP"按鈕，將程式存檔並寫入到 NodeMCU 後執行。

4. 實驗成果：

(1) 按下鍵盤就放會印出按下按鍵的值，終端機顯示的變化如圖 5-13。

```
Key[1][1]=1
Key[1][2]=2
Key[1][3]=3
Key[1][4]=A
Key[2][1]=4
Key[2][2]=5
Key[2][3]=6
Key[2][4]=B
Key[3][1]=7
Key[3][2]=8
Key[3][3]=9
Key[3][4]=C
Key[4][1]=*
Key[4][2]=0
Key[4][3]=#
Key[4][4]=D
```

圖 5-13

```
> dofile("Lab5-B.lua");
> Key[1][1]=1
Key[1][1]=1
Key[1][1]=1
Key[1][1]=1
Key[1][1]=1
Key[1][1]=1
Key[1][1]=1
Key[1][1]=1
Key[1][1]=1
Key[1][1]=1
Key[1][1]=1
Key[1][1]=1
Key[1][1]=1
Key[1][1]=1
Key[1][1]=1
```

```
Key[1][2]=2
Key[1][2]=2
Key[1][2]=2
Key[1][2]=2
Key[1][2]=2
Key[1][2]=2
Key[1][2]=2
Key[1][2]=2
Key[1][2]=2
Key[1][4]=A
Key[1][4]=A
Key[1][4]=A
Key[1][4]=A
Key[1][4]=A
Key[1][4]=A
```

圖 5-14

(2) 長時間按下鍵盤才放，終端機顯示的變化如圖 5-14，當按下按鍵不放時每 0.2 秒會印出按下按鍵的值，直到手放開為止。

61

主題 C：使用 Fritzing、Snap4NodeMCU 與 ESPlorer IDE 平台做鍵盤感測器結合蜂鳴器之軟硬體設計

1. 題目：使用 ESPlorer IDE 平台撰寫讀取鍵盤感測器的值，根據按下按鍵的不同，調整蜂鳴器的聲音。

2. 實驗目標：以蜂鳴器發出聲音來確認已經有按下鍵盤感測器上的按鍵，並利用 PWM 原理進一步的改變蜂鳴器發出的聲音。

3. 實驗步驟：此實驗將分三個階段進行，

    a. 使用 Fritzing 實際產生麵包板電路圖、電路概要圖以及實際的麵包板接線圖。

    b. 加入蜂鳴器，調整 Lua 程式，當按下一個鍵之後，自動嗶一聲來提醒已經按下鍵盤上的按鍵。

    c. 調整 Lua 程式，運用 PWM，改變蜂鳴器聲音。

a. 使用 Fritzing 實際產生麵包板電路圖、電路概要圖以及實際的麵包板接線圖

    (1) 延續主題 B，搜尋"buzzer"，將蜂鳴器拉到麵包板上，如圖 5-15。

圖 5-15

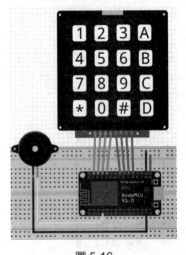
圖 5-16

    (2) 將蜂鳴器的正極接腳接到 NodeMCU 上的 D12(SD3)數位接孔，負極接腳接到 NodeMCU 上的 GND，如圖 5-16。

(3) 點選上方的"概要圖"標籤即可看到自動幫我們建好的電路概要圖,再將元件以及線路擺放至較美觀的位置,並將線路佈線完成,如圖 5-17。

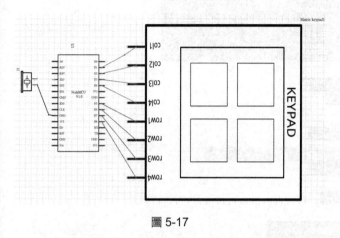

圖 5-17

圖 5-18

(4) 接續主題 B 的實際麵包板接線,再將蜂鳴器的正極接腳接到 NodeMCU 的 D12 接腳,負極接腳接到 NodeMCU 上的 GND 接孔,如圖 5-18。

b. 加入蜂鳴器,調整 Lua 程式,當按下一個鍵之後,自動嗶一聲來提醒已經按下鍵盤上的按鍵

(1) 點擊 ESPlorer.bat 工具,進入開發環境 IDE。在左方的程式撰寫區撰寫 Lua 程式,設定蜂鳴器接腳為 12,一開始要先將電位設為低電位,否則一接好線路時就會發出聲響,如圖 5-19(a)。也可以使用 Snap4NodeMCU 加以實現,如圖 5-19(b)。

```lua
1   buzzer=12                          -- set buzzer pin to 12
2   gpio.mode(buzzer,gpio.OUTPUT)      -- set pin 12 as output pin
3   gpio.write(buzzer,gpio.LOW)        -- set pin 12 to low
4   -- define 4x4 keypad name
5   keys={{"1","2","3","A"},{"4","5","6","B"},{"7","8","9","C"},{"*","0","#","D"}}
6   for i=1,8 do
7     gpio.mode(i,gpio.OUTPUT)         -- set pin 1~4 as output pin
8     gpio.write(i,gpio.LOW)           -- set pin 1~8 to low
9   end
10  for i=5,8 do
11    gpio.mode(i,gpio.INPUT)          -- set pin 5~8 as input pin
12  end
13  tmr.alarm(0,200,tmr.ALARM_AUTO,function()  -- repeat very 0.2 second
14    gpio.write(buzzer,gpio.LOW)      -- set pin 12 to low (disable buzzer)
15    -- key scan algorithm
16    for r=1,4 do                     -- loop for each row pin
17      gpio.write(r,gpio.HIGH)        -- set row pin to high
18      for c=1,4 do                   -- loop for each column pin
19        dValue=gpio.read((c + 4))    -- read value of column pin
20        if (dValue == 1) then        -- if the value is 1 then a key is pressed
21          print("r=",r,"c=",c,"key=",keys[r][c])  -- print out which key
22          gpio.write(buzzer,gpio.HIGH)            -- set pin 12 to high (enable buzzer)
23        end
24      end
25      gpio.write(r,gpio.LOW)         -- set row pin to low
26    end
27  end)
```

圖 5-19(a)

63

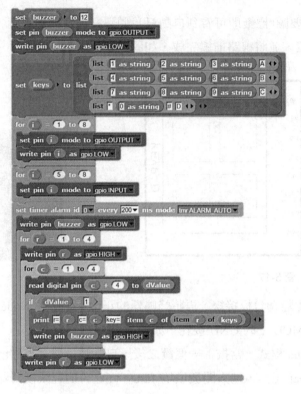

圖 5-19(b)

(2) 實際麵包板接線方式如同步驟 a，詳細接線圖如圖 5-20。

圖 5-20

(3) 到 ESPlorer IDE 平台點擊左下方"Save to ESP"按鈕，將程式存檔並寫入到 NodeMCU 後執行。

c. 調整 Lua 程式，運用 PWM，改變蜂鳴器聲音

   (1) 修改步驟 b 的程式碼，設定變數 width 爲 PWM 音量大小頻率，可以自行設定讓每一按鍵的頻率都不相同來發出不同的聲音，程式碼範例如圖 5-21。

```
1   keys = {{"1","2","3","A"},{"4","5","6","B"},{"7","8","9","C"},{"*","0","#","D"}}
2                                        -- define keys' name for 4x4
3   buzzerpin = 12
4   gpio.mode(buzzerpin, gpio.OUTPUT)
5   gpio.write(buzzerpin, gpio.LOW)      --set buzzerpin to Low
6
7   for pin = 1, 4 do
8   gpio.mode(pin, gpio.OUTPUT)          --initial gpio D1~D4 output mode
9   gpio.mode(pin+4, gpio.OUTPUT)        --initial gpio D5~D8 output mode then set to Low
10  gpio.write(pin+4, gpio.LOW)
11  gpio.mode(pin+4, gpio.INPUT)         --Set pin D5~D8 input mode (Row)
12  end
13  tmr.alarm(0,200,1,function()         --Set Timer in 200ms
14      for rpin =1,4 do                 --Loop for Set high level to assigned row pin
15          gpio.write(rpin,gpio.HIGH)
16          for cpin=1,4 do              --Loop for read assignedd column pin for keyin detect
17          hit = gpio.read(cpin+4)
18              if hit == gpio.HIGH then
19                  width = rpin*200 + cpin*50
20
21                  gpio.write(buzzerpin, gpio.HIGH)
22
23                  pwm.setup(buzzerpin,width,width/2)
24                  pwm.start(buzzerpin)
25                  pwm.stop(buzzerpin)
26
27                  print ("Key["..rpin.."]["..cpin.."]="..keys[rpin][cpin] .. ", width=" ..width)
28                          -- If there is keyin , show Row/Col pin name and Key name
29              end
30          end
31          gpio.write(rpin,gpio.LOW)    --reset Row to low level
32      end
33  end)
```

圖 5-21

   (2) 到 ESPlorer IDE 平台點擊左下方"Save to ESP"按鈕，將程式存檔並寫入到 NodeMCU 後執行。

4. 實驗成果：

a. 使用 Fritzing 實際產生麵包板電路圖、電路概要圖以及 PCB 印刷板電路圖，以及實際的麵包板接線圖

   (1) 麵包板電路圖如圖 5-22。

圖 5-22

(2) 電路概要圖如圖 5-23。

(3) 實際的麵包板接線圖如圖 5-24。

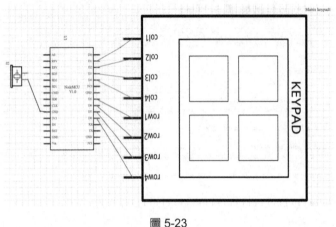

圖 5-23

圖 5-24

b. 加入蜂鳴器，調整 Lua 程式，當按下一個鍵之後，自動嗶一聲來提醒已經按下鍵盤上的按鍵

當使用者線路接好，撰寫好程式後執行，當在按下按鍵時蜂鳴器就會發出聲響讓使用者知道已經有按下按鍵。

c. 調整 Lua 程式，運用 PWM，改變蜂鳴器聲音

藉由設定每個按鍵之頻率值的不同，因此按下每個按鍵所發出的聲音也會不同，終端機顯示結果如圖 5-25。

```
Key[1][1]=1, width=250
Key[1][2]=2, width=300
Key[1][3]=3, width=350
Key[1][4]=A, width=400
Key[2][1]=4, width=450
Key[2][2]=5, width=500
Key[2][3]=6, width=550
Key[2][4]=B, width=600
Key[3][1]=7, width=650
Key[3][2]=8, width=700
Key[3][3]=9, width=750
Key[3][4]=C, width=800
Key[4][1]=*, width=850
Key[4][2]=0, width=900
Key[4][3]=#, width=950
Key[4][4]=D, width=1000
```

圖 5-25

主題 A：使用 Fritzing IDE 平台做土壤濕度感測器電路設計

1. 題目：使用 Fritzing 工具畫出 NodeMCU 與土壤濕度感測器連接電路。

2. 實驗目標：使用土壤濕度感測器範例練習 Fritzing 的電路設計，實際產生麵包板電路圖、電路概要圖以及 PCB 印刷板電路圖。

3. 實驗步驟：此實驗將分三個階段進行，

   a. 產生麵包板電路圖。

   b. 產生電路概要圖。

   c. 產生 PCB 印刷板電路圖。

a. 產生麵包板電路圖

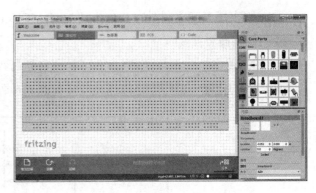

   (1) 開啓 Fritzing 工具，點選"麵包板"，如圖 6-1。

   (2) 在右方搜尋窗格輸入"NodeMCU"，將 NodeMCU 拖曳至麵包板上並完成接線，如圖 6-2。

   (3) 搜尋元件"soil"，將土壤濕度感測器拖曳至麵包板上，如圖 6-3。

圖 6-1

圖 6-2

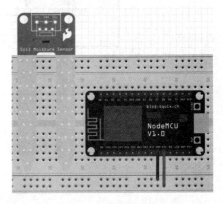

圖 6-3

註：由於 Fritzing 目前尚未提供 YL-69 土壤濕度感測模組的元件，因此使用 SparkFun Soil Moisture Sensor 為例來建立電路圖。

(4) 將土壤濕度感測器的 VCC 和 GND 分別接到麵包板上的電源和接地接孔，如圖 6-4。

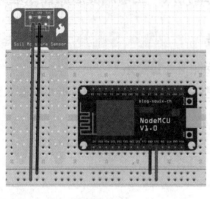
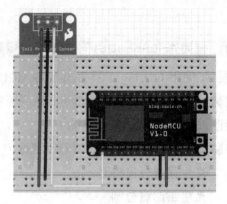

圖 6-4　　　　　　　　　　　　　　　　圖 6-5

(5) 將土壤濕度感測器的 SIG 接到 NodeMCU 上的 A0 接腳，如圖 6-5。

b. 產生電路概要圖

　　點選上方的"概要圖"標籤即可看到自動幫我們建好的電路概要圖。將元件以及線路擺放至較美觀的位置，並將線路佈線完成，如圖 6-6 為修正後的電路概要圖。

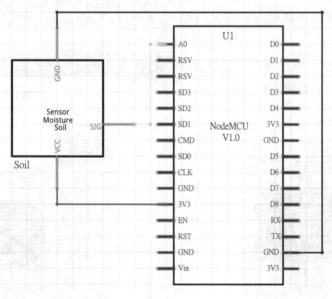

圖 6-6

c. 產生 PCB 印刷板電路圖

　　點選上方的"PCB"標籤看到自動幫我們建好
的 PCB 印刷板電路圖。將元件以及線路擺
放至較美觀的位置，並將線路佈線完成，如
圖 6-7 為修正後的 PCB 印刷版電路圖。

4. 實驗成果：

　　(1)　麵包板電路如圖 6-8。

　　(2)　電路概要圖如圖 6-9。

　　(3)　PCB 印刷板電路圖如圖 6-10。

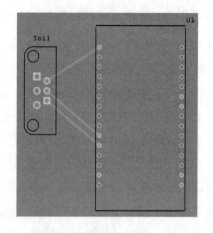

圖 6-7

圖 6-8

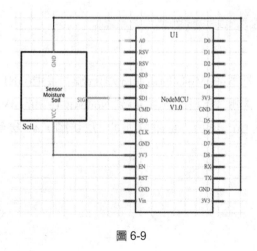

圖 6-9

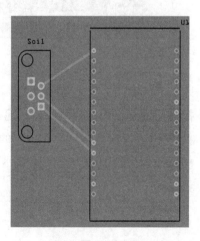

圖 6-10

主題 B：使用 Snap4NodeMCU 與 ESPlorer IDE 平台做土壤濕度感測類比模組程式設計

1. 題目：實際利用麵包板將 NodeMCU 以及土壤濕度感測模組進行線路連接，並使用 ESPlorer IDE 平台撰寫讀取土壤濕度感測模組的類比值，並觀察以及實驗下述四點：

   (1) 將土壤濕度感測模組置入水杯，觀察終端機顯示的變化。

   (2) 將土壤濕度感測模組移出水杯，觀察終端機顯示的變化。

   (3) 將土壤濕度感測模組置入水杯後，慢慢移出水杯，觀察終端機顯示的變化。

   (4) 調整可變電阻，觀察終端機顯示的變化。

2. 實驗目標：使用讀取土壤濕度感測模組的類比值程式練習 Lua 的程式設計，並實際觀察麵包板接線完成後，程式進行時 ESPlorer 終端機顯示畫面，讓使用者實際了解當土壤濕度感測模組有水和沒水時，實際讀取類比值的變化情形。

3. 實驗步驟：

   (1) 點擊 ESPlorer.bat 工具，進入開發環境 IDE。在左方的程式撰寫區撰寫 Lua 程式，每 1 秒呼叫函數傳回土壤濕度的讀取值，如圖 6-11(a)。也可以使用 Snap4NodeMCU 加以實現，如圖 6-11(b)。

```
1  □tmr.alarm(0,1000,tmr.ALARM_AUTO,function()  -- repeat every 1 second
2      aValue=adc.read(0)                       -- read analog value from pin 0
3      print("Soil moisture=",aValue)           -- print out Soil moisture
4    └end)
```

圖 6-11(a)

圖 6-11(b)

   (2) 將 NodeMCU 插入到麵包板上，再利用杜邦線把 NodeMCU 的電源(3V3)接腳和接地(GND)接腳接到麵包板上的電源和接地接孔。將土壤感測器模組的 VCC 接到 NodeMCU 上的 3V3 接腳，GND 接到 NodeMCU 上的 GND 接腳，D0 接到 NodeMCU 上的 D2 接腳，A0 接到 NodeMCU 上的 A0 接腳，如圖 6-12。

圖 6-12

(3) 到 ESPlorer IDE 平台點擊左下方"Save to ESP"按鈕，將程式存檔並寫入到 NodeMCU 後執行。

4. 實驗成果：

(1) 將土壤濕度感測模組置入水杯，終端機顯示的變化如圖 6-13。虛線下方式將感測模組置入水中後的類比值變化，大約放置五分鐘左右類比值大約在 550 左右跳動。

			analog value=529
			analog value=528
			analog value=531
			analog value=530
analog value=1021			analog value=533
analog value=1024			analog value=532
analog value=1023			analog value=536
analog value=1024			analog value=537
analog value=772	analog value=556		analog value=538
analog value=405	analog value=553		analog value=539
analog value=403	analog value=556	analog value=786	analog value=539
analog value=403	analog value=556	analog value=1024	analog value=540
analog value=402	analog value=554	analog value=1021	analog value=544
analog value=401	analog value=556	analog value=1021	analog value=548
analog value=400	analog value=553	analog value=1021	analog value=608
analog value=402	analog value=553	analog value=1024	analog value=587
analog value=403	analog value=554	analog value=1021	analog value=631
analog value=404	analog value=554	analog value=1021	analog value=634
analog value=404	analog value=554	analog value=1024	analog value=729
analog value=406	analog value=553	analog value=1024	analog value=786
analog value=407	analog value=554	analog value=1024	analog value=1024
analog value=409	analog value=556	analog value=1024	analog value=1021
analog value=425		analog value=1024	analog value=1021
analog value=426		analog value=1024	analog value=1021
		analog value=1024	analog value=1024

圖 6-13   圖 6-14   圖 6-15

(2) 將土壤濕度感測模組移出水杯，終端機顯示的變化如圖 6-14。將感測模組移出水杯後，類比值就會回到 1024。

(3) 將土壤濕度感測模組置入水杯後，慢慢移出水杯，終端機顯示的變化如圖 6-15。可以發現當慢慢地將感測模組移出水杯時，類比值會漸漸地變大。

(4) 調整可變電阻(圖 6-16)，終端機顯示的變化如圖 6-17。調整可變電阻可為改變數位接腳的門檻值，對類比接腳無直接影響。

圖 6-16

```
analog value=1024        analog value=1024
analog value=1024        analog value=1021
analog value=1024        analog value=1024
analog value=1024        analog value=1021
analog value=952         analog value=1021
analog value=929         analog value=603
analog value=700         analog value=551
analog value=602         analog value=508
analog value=596         analog value=1024
analog value=1024        analog value=1024
analog value=1024        analog value=1024
analog value=1024        analog value=1024
```

可變電阻往左調到底   可變電阻往右調到底

圖 6-17

主題 C：使用 Snap4NodeMCU 與 ESPlorer IDE 平台做土壤濕度感測數位模組程式設計

1. 題目：實際利用麵包板將 NodeMCU 以及土壤濕度感測模組進行線路連接，並使用 ESPlorer IDE 平台撰寫讀取土壤濕度感測模組的數位值，並觀察以及實驗下述三點：
   (1) 將土壤濕度感測模組置入水杯，觀察終端機顯示的變化。
   (2) 將土壤濕度感測模組移出水杯，觀察終端機顯示的變化。
   (3) 調整可變電阻，觀察終端機顯示的變化。

2. 實驗目標：使用讀取土壤濕度感測模組的數位值程式練習 Lua 的程式設計，並實際觀察麵包板接線完成後，程式進行時 ESPlorer 終端機顯示畫面，讓使用者實際了解當土壤濕度感測模組有水和沒水時，實際讀取數位值的變化情形。

3. 實驗步驟：
   (1) 點擊 ESPlorer.bat 工具，進入開發環境 IDE。在左方的程式撰寫區撰寫 Lua 程式，每 1 秒呼叫函數傳回土壤濕度的讀取值，如圖 6-18(a)。也可以使用 Snap4NodeMCU 加以實現，如圖 6-18(b)。
   (2) 土壤感測器模組的實際電路板接法和主題 B 相同，如圖 6-19。
   (3) 到 ESPlorer IDE 平台點擊左下方"Save to ESP"按鈕，將程式存檔並寫入到 NodeMCU 後執行。

```
1   soil=2                                          -- set soil pin to pin 2
2   gpio.mode(soil,gpio.INPUT)                      -- set pin 2 as input pin
3   tmr.alarm(0,1000,tmr.ALARM_AUTO,function()      -- repeat every 1 second
4     dValue=gpio.read(soil)                        -- read digital pin 2
5     print("Soil moisture=",dValue)                -- print out Soil moisture
6   end)
7   end)
```

圖 6-18(a)

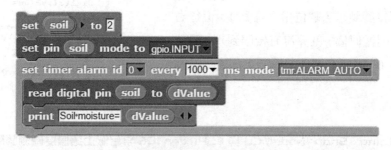

圖 6-18(b)

圖 6-19

4. 實驗成果:

(1) 將土壤濕度感測模組置入水杯,終端機顯示的變化如圖 6-20。當有水時數位讀取值為 0。

圖 6-20

73

(2) 將土壤濕度感測模組移出水杯,終端機顯示的變化如圖 6-21。將感測模組移出水杯後,數位值為 1。

(3) 調整可變電阻可為改變數位接腳的門檻值,調整可變電阻時,當往左調到底時,因為門檻值高,就算接觸到水,數位值一樣為 1;但往右調底時,門檻值低,就算沒有接觸到水,數位值一樣為 0,因此在這裡門檻值太高或太低都不好。

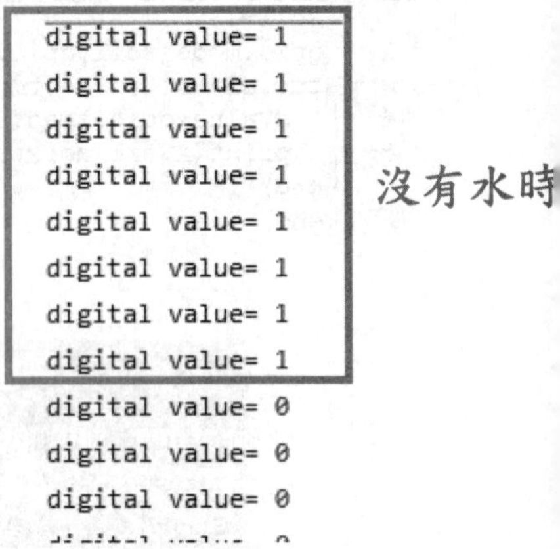

```
digital value= 1
digital value= 1
digital value= 1
digital value= 1
digital value= 1
digital value= 1
digital value= 1
digital value= 1
digital value= 0
digital value= 0
digital value= 0
```

沒有水時

圖 6-21

主題 D:使用 Fritzing、Snap4NodeMCU 與 ESPlorer IDE 平台做土壤濕度感測模組結合蜂鳴器 (或 LED) 之軟硬體設計

1. 題目:使用 ESPlorer IDE 平台撰寫讀取土壤濕度感測模組的值,當快缺水時,自動一直嗶聲(或開啟 LED 燈)來提醒擁有者。

2. 實驗目標:以蜂鳴器發出聲音或是開啟 LED 燈來確認土壤是否有缺水。

3. 實驗步驟:此實驗將分三個階段進行,

   a. 使用 Fritzing 實際產生麵包板電路圖、電路概要圖以及 PCB 印刷板電路圖,以及實際的麵包板接線圖。

   b. 加入蜂鳴器(或 LED),調整 Lua 程式,當快缺水時,自動一直嗶聲(或開啟 LED 燈)來提醒擁有者。

a. 使用 Fritzing 實際產生麵包板電路圖、電路概要圖以及 PCB 印刷板電路圖,以及實際的麵包板接線圖

   (1) 延續主題 B,搜尋"led"和"resistor",將 LED 燈和電阻拉到麵包板上,其中電阻要和 LED 的短邊接腳相接,如圖 6-22。

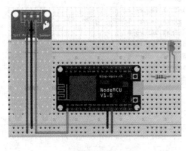

圖 6-22

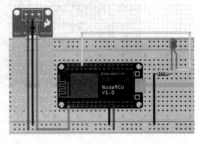

圖 6-23

(2) 將 LED 燈的長邊接腳接到 NodeMCU 上的 D3 數位接孔，電阻的另一端接腳接到麵包板上的接地接孔，如圖 6-23。

(3) 點選上方的"概要圖"標籤即可看到自動幫我們建好的電路概要圖，將元件以及線路擺放至較美觀的位置，並將線路佈線完成，如圖 6-24。

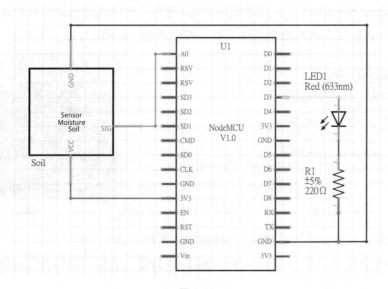

圖 6-24

(4) 點選上方的"PCB"標籤看到自動幫我們建好的 PCB 印刷板電路圖，將元件以及線路擺放至較美觀的位置，並將線路佈線完成，如圖 6-25。

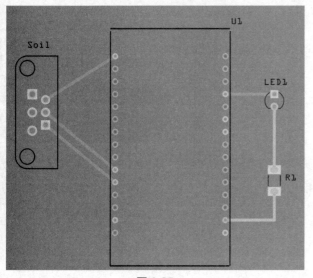

圖 6-25

圖 6-26

(5) 接續主題 B 的實際麵包板接線，再將 LED 的長邊接腳接到 NodeMCU 的 D3 接腳，短邊接腳和電阻相接，而電阻的另一端接腳接到麵包板上的接地接孔，如圖 6-26。

b. 加入蜂鳴器(或 LED)，調整 Lua 程式，當快缺水時，自動一直嗶聲(或開啟 LED 燈)來提醒擁有者

(1) 點擊 ESPlorer.bat 工具，進入開發環境 IDE。在左方的程式撰寫區撰寫 Lua 程式，設定 LED 接腳為 3，再經由程式判斷是否缺水而亮燈或關燈，如圖 6-27。

```
1  analogpin = 0  --set analog pin is 0
2  ledpin = 3
3  gpio.mode(ledpin,gpio.OUTPUT)
4  tmr.alarm(0, 1000, 1, function()
5      watervalue = adc.read(analogpin)
6      if watervalue > 700 then
7          gpio.write(ledpin, gpio.HIGH)
8          print("analog value=" ..watervalue ..",Light ON")
9      else
10         gpio.write(ledpin, gpio.LOW)
11         print("analog value=" ..watervalue ..",Light OFF")
12     end
13 end)
```

圖 6-27

圖 6-28

(2) 實際麵包板接線方式如同步驟 a，詳細接線圖如圖 6-28。

(3) 到 ESPlorer IDE 平台點擊左下方"Save to ESP"按鈕，將程式存檔並寫入到 NodeMCU 後執行。

4. 實驗成果：

a. 使用 Fritzing 實際產生麵包板電路圖、電路概要圖以及 PCB 印刷板電路圖，以及實際的麵包板接線圖

(1) 麵包板電路圖如圖 6-29。

(2) 電路概要圖如圖 6-30。

(3) PCB 印刷板電路圖如圖 6-31。

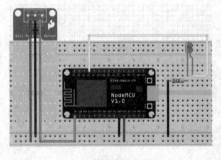

圖 6-29

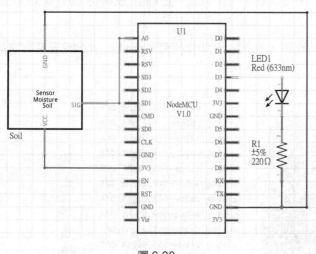

圖 6-30

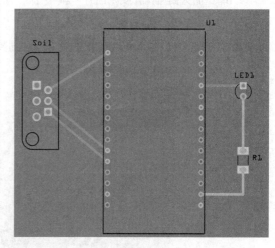

圖 6-31

(4) 實際的麵包板接線圖如圖 6-32。

圖 6-32

77

b. 加入蜂鳴器(或 LED)，調整 Lua 程式，當快缺水時，自動一直嗶聲(或開啓 LED 燈)來提醒擁有者

設定當土壤感測模組的數位值高於 700 時，表示缺水即會亮燈；而低於 700 時表示土壤濕度足夠就會把燈關掉，如圖 6-33。

```
analog value=513,Light OFF
analog value=524,Light OFF
analog value=529,Light OFF
analog value=534,Light OFF
analog value=541,Light OFF
analog value=544,Light OFF
analog value=548,Light OFF
analog value=549,Light OFF
analog value=551,Light OFF
analog value=555,Light OFF
analog value=554,Light OFF
analog value=557,Light OFF
analog value=560,Light OFF
analog value=563,Light OFF
analog value=984,Light ON
analog value=1024,Light ON
analog value=1021,Light ON
analog value=1024,Light ON
analog value=1021,Light ON
analog value=1021,Light ON
analog value=1024,Light ON
analog value=1021,Light ON
analog value=1024,Light ON
analog value=1021,Light ON
analog value=1024,Light ON
analog value=1021,Light ON
analog value=1024,Light ON
analog value=1021,Light ON
analog value=1021,Light ON
```

飽水時暗燈

缺水時亮燈

圖 6-33

主題 A：使用 Fritzing IDE 平台做溫濕度感測器電路設計

1. 題目：使用 Fritzing 工具畫出 NodeMCU 與溫濕度感測器連接電路。

2. 實驗目標：使用溫濕度感測器範例練習 Fritzing 的電路設計，實際產生麵包板電路圖、電路概要圖以及 PCB 印刷板電路圖。

3. 實驗步驟：此實驗將分三個階段進行，

   a. 產生麵包板電路圖。

   b. 產生電路概要圖。

   c. 產生 PCB 印刷板電路圖。

a. 產生麵包板電路圖

   (1) 開啟 Fritzing 工具，點選"麵包板"，如圖 7-1。

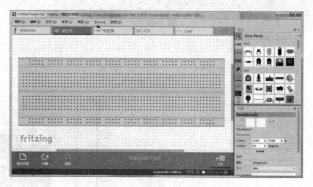

圖 7-1

   (2) 在右方搜尋窗格輸入"NodeMCU"，將 NodeMCU 拖曳至麵包板上並完成接線，如圖 7-2。

   (3) 搜尋元件"RHT"，將溫濕度感測器拖曳至麵包板上，如圖 7-3。由於接線關係，可以利用點選元件按右鍵選擇旋轉，將元件旋轉 180°，如圖 7-4。旋轉後的圖如圖 7-5。此外點選感測器上接腳，會解釋該接腳的功能為何，以圖 7-6 為例，該接腳為 VCC，用來接電源。

圖 7-2

圖 7-3

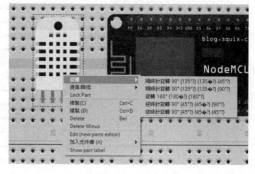

圖 7-4

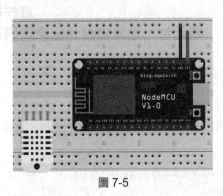
圖 7-5

圖 7-6

(4) 將溫濕度感測器的 VCC 以及 GND 接腳分別接到麵包板上的電源和接地接孔，Data 接腳接到 NodeMCU 上的 D1 數位接孔，如圖 7-7。

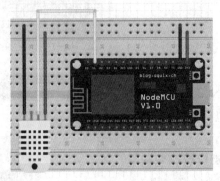
圖 7-7

b. 產生電路概要圖

點選上方的"概要圖"，將元件以及線路擺放至較美觀的位置，並將線路佈線完成，如圖 7-8。

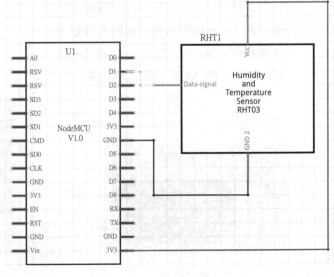
圖 7-8

80

c. 產生 PCB 印刷板電路圖

　　點選上方的"PCB"，將元件以及線路擺放至較美觀的位置，並將線路佈線完成，如圖 7-9。

4. 實驗成果：

(1) 麵包板電路如圖 7-10。

(2) 電路概要圖如圖 7-11。

(3) PCB 印刷板電路圖如圖 7-12。

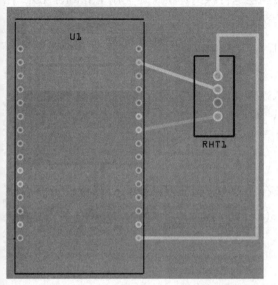

圖 7-9

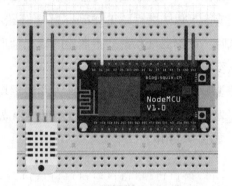

圖 7-10

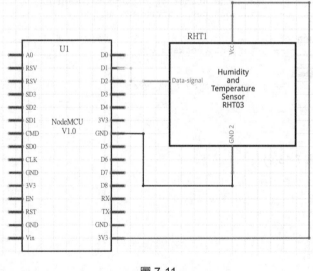

圖 7-11

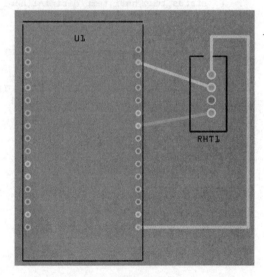

圖 7-12

81

主題 B：使用 Snap4NodeMCU 與 ESPlorer IDE 平台做溫濕度感測器程式設計

1. 題目：實際利用麵包板將 NodeMCU 以及溫濕度感測器進行線路連接，並使用 ESPlorer IDE 平台撰寫讀取溫濕度感測器的值，並觀察以及實驗下述兩點：

    (1) 手輕輕按住感測模組，觀察終端機顯示的變化。

    (2) 對著感測模組吹氣，觀察終端機顯示的變化。

2. 實驗目標：使用讀取溫濕度感測器模組之範例程式練習 Lua 的程式設計，並實際觀察麵包板接線完成後，程式進行時 ESPlorer 終端機顯示畫面，讓使用者實際了解對溫濕度感測模組蓋住或吹氣時，實際溫度和濕度讀取值的變化情形。

3. 實驗步驟：

    (1) 點擊 ESPlorer.bat 工具，進入開發環境 IDE。在左方的程式撰寫區撰寫 Lua 程式，每 2 秒呼叫函數傳回溫度和濕度的讀取值，如圖 7-13(a)。也可以使用 Snap4NodeMCU 加以實現，如圖 7-13(b)。

```
1   gpio.mode(7,gpio.INPUT)                          -- set pin 7 as input
2   tmr.alarm(0,2000,tmr.ALARM_AUTO,function()        -- repeat every 2 seconds
3       status,temp,humi,temp_decimial,humi_decimial=dht.read(7)   -- read temperature and humidity
4       if (status == dht.OK) then                    -- check the status of reading DHT sensor
5           print("t=",temp,"h=",humi)                -- print out temperature and humidity
6       else
7           if (status == dht.ERROR_CHECKSUM) then
8               print("DHT Checksum error.")           -- print out Checksum error
9           else
10              if (status == dht.ERROR_TIMEOUT) then
11                  print("DHT Time Out error.")       -- print out Time Out error
12              end
13          end
14      end
15  end)
```

圖 7-13(a)

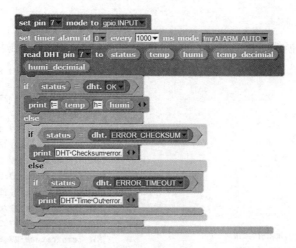

圖 7-13(b)

(2) 將 NodeMCU 插入到麵包板上,再利用杜邦線把 NodeMCU 的電源(3V3)接腳和接地(GND)
接腳接到麵包板上的電源和接地接孔。將溫濕度感測模組的"+"接到 NodeMCU 上的 3V3
接孔,"-"接到 NodeMCU 上的 GND 接孔,"OUT"接到 NodeMCU 上的 D1 數位接孔,如圖
7-14。

(3) 到 ESPlorer IDE 平台點擊左下方"Save to ESP"按鈕,將程式存檔並寫入到 NodeMCU 後執
行。

圖 7-14

4. 實驗成果：

(1) 手輕輕按住感測模組，終端機顯示的變化如圖 7-15。虛線下方為將感測模組蓋住時的變化情形，可以發現除了溫度有慢慢提升之外，因為人手也是有濕氣的，所以濕度也有跟著改變。

```
DHT Temperature:26.7; Humidity:35
DHT Temperature:26.7; Humidity:35
DHT Temperature:26.7; Humidity:35
DHT Temperature:26.7; Humidity:35
DHT Temperature:26.7; Humidity:39
DHT Temperature:26.7; Humidity:49.3
DHT Temperature:26.8; Humidity:57.6
DHT Temperature:26.8; Humidity:62.6
DHT Temperature:26.8; Humidity:66
DHT Temperature:26.9; Humidity:66
DHT Temperature:26.9; Humidity:68.2
DHT Temperature:27; Humidity:70.5
DHT Temperature:27; Humidity:72.3
DHT Temperature:27.1; Humidity:73.8
DHT Temperature:27.2; Humidity:74
DHT Temperature:27.3; Humidity:75.5
DHT Temperature:27.3; Humidity:76.2
```

圖 7-15

```
> DHT Temperature:27.8; Humidity:32.4
DHT Temperature:27.8; Humidity:32.5
DHT Temperature:27.8; Humidity:32.5
DHT Temperature:27.8; Humidity:32.5
DHT Temperature:27.8; Humidity:32.6
DHT Temperature:27.8; Humidity:32.5
DHT Temperature:27.8; Humidity:32.3
DHT Temperature:27.8; Humidity:32.3
DHT Temperature:27.8; Humidity:32.3
DHT Temperature:27.8; Humidity:32.3
DHT Temperature:28.9; Humidity:77.2
DHT Temperature:29.1; Humidity:87.5
DHT Temperature:28.8; Humidity:89.1
DHT Temperature:28.2; Humidity:89
DHT Temperature:27.8; Humidity:87
DHT Temperature:27.8; Humidity:85.3
DHT Temperature:27.8; Humidity:79
DHT Temperature:27.8; Humidity:66.5
DHT Temperature:27.8; Humidity:53.2
DHT Temperature:27.9; Humidity:45.5
DHT Temperature:27.9; Humidity:41.5
DHT Temperature:27.9; Humidity:38.8
DHT Temperature:27.9; Humidity:37.7
DHT Temperature:27.9; Humidity:35.8
```

圖 7-16

(2) 對著感測模組吹氣，終端機顯示的變化如圖 7-16。虛線下方為對感測模組吹氣時的變化情形，可以發現除了溫度有慢慢提升之外，濕度也有大幅改變；此外溫度下降的速度比濕度下降的快。

主題 C：使用 Fritzing、Snap4NodeMCU 與 ESPlorer IDE 平台做溫濕度感測器結合土壤濕度感測器之軟硬體設計

1. 題目：使用 ESPlorer IDE 平台撰寫讀取溫濕度感測器以及土壤濕度感測器的值。

2. 實驗目標：讓使用者利用 Lua 程式範例練習，同時讀取環境的溫度和濕度以及土壤的濕度值。

3. 實驗步驟：此實驗將分兩個階段進行，

    a. 使用 Fritzing 實際產生麵包板電路圖、電路概要圖以及 PCB 印刷板電路圖，以及實際的麵包板接線圖。

    b. 加入土壤濕度感測器，調整 Lua 程式，同時顯示植物生長時候的溫濕度及土壤濕度。

a. 使用 Fritzing 實際產生麵包板電路圖、電路概要圖以及 PCB 印刷板電路圖，以及實際的麵包板接線圖

(1) 延續主題 B，搜尋"soil"，將土壤濕度感測器拉到麵包板上，且為了線路美觀將 NodeMCU 的另一邊 VCC 和 GND 也接到麵包板上，如圖 7-17。

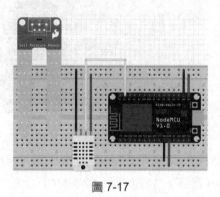

圖 7-17　　　　　　　　　　　　　　　　圖 7-18

(2) 將土壤濕度感測器的 VCC 和 GND 分別接到麵包板上的電源和接地接孔，土壤濕度感測器的 SIG 接到 NodeMCU 上的 A0 接腳，如圖 7-18。

(3) 點選上方的"概要圖"，將元件以及線路擺放至較美觀的位置，並將線路佈線完成，如圖 7-19。

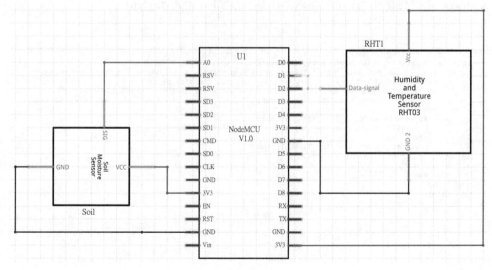

圖 7-19

(4) 點選上方的"PCB"，將元件以及線路擺放至較美觀的位置，並將線路佈線完成，如圖 7-20。

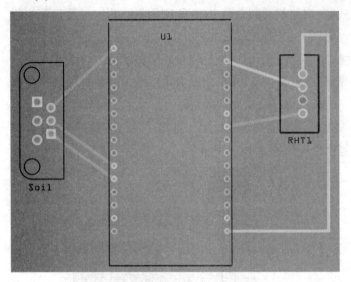

圖 7-20

圖 7-21

(5) 接續主題 B 的實際麵包板接線，再將土壤濕度感測模組的 VCC 和 GND 分別接到 NodeMCU 的 3V3 和 GND 接腳，A0 接到 NodeMCU 上的 A0 接腳，D0 腳接到 NodeMCU 上的 D2 接腳(D0 在本實驗主題中可不接)，如圖 7-21。

b. 加入土壤濕度感測器，調整 Lua 程式，同時顯示植物生長時候的溫濕度及土壤濕度

(1) 點擊 ESPlorer.bat 工具，進入開發環境 IDE。在左方的程式撰寫區撰寫 Lua 程式，如圖 7-22(a)。也可以使用 Snap4NodeMCU 加以實現，如圖 7-22(b)。

```lua
1    gpio.mode(7,gpio.INPUT)                      -- set pin 7 as input
2    tmr.alarm(0,2000,tmr.ALARM_AUTO,function()   -- repeat every 2 seconds
3      status,temp,humi,temp_decimial,humi_decimial=dht.read(7)  -- read temperature and humidity
4      soil=adc.read(0)                           -- read humidity of soil
5      if (status == dht.OK) then                 -- check the status of reading DHT sensor
6        print("t=",temp,"h=",humi,"soil=",soil)              -- print out temperature and humidity
7      else
8        if (status == dht.ERROR_CHECKSUM) then
9          print("DHT Checksum error.")           -- print out Checksum error
10       else
11         if (status == dht.ERROR_TIMEOUT) then
12           print("DHT Time Out error.")         -- print out Time Out error
13         end
14       end
15     end
16   end)
```

圖 7-22(a)

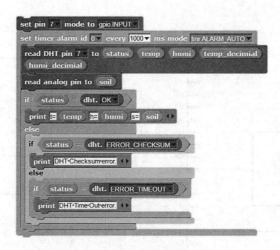

圖 7-22(b)

(2) 實際麵包板接線方式如同步驟 a，詳細接線圖如圖 7-23。

圖 7-23

(3) 到 ESPlorer IDE 平台點擊左下方"Save to ESP"按鈕，將程式存檔並寫入到 NodeMCU 後執行。

4. 實驗成果：

a. 使用 Fritzing 實際產生麵包板電路圖、電路概要圖以及 PCB 印刷板電路圖，以及實際的麵包板接線圖

　(1) 麵包板電路圖如圖 7-24。

　(2) 電路概要圖如圖 7-25。

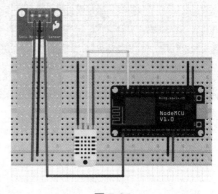

圖 7-24

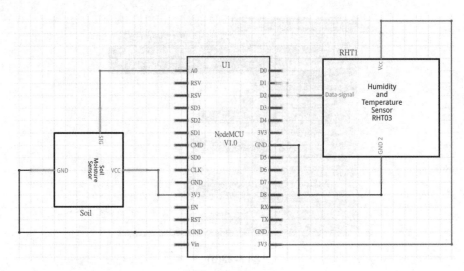

圖 7-25

(3) PCB 印刷板電路圖如圖 7-26。

(4) 實際的麵包板接線圖如圖 7-27。

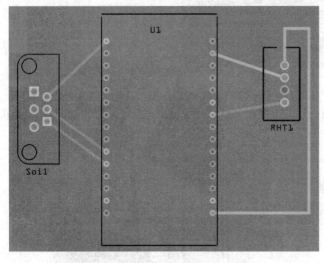

圖 7-26

圖 7-27

b. 加入土壤濕度感測器，調整 Lua 程式，同時顯示植物生長時候的溫濕度及土壤濕度

終端機顯示結果如圖 7-28。虛線上方是當土壤濕度感測模組沒有偵測到水時的讀取值，虛線下方則是將土壤濕度感測模組放入水中後的讀取值。

圖 7-28

主題 D：使用 Fritzing、Snap4NodeMCU 與 ESPlorer IDE 平台做溫濕度感測器結合土壤濕度感測器以及蜂鳴器之軟硬體設計

1. 題目：使用 ESPlorer IDE 平台撰寫讀取溫濕度感測器以及土壤濕度感測器的值，當快缺水時，溫度過高或是過低時，濕度過高或是過低時，自動一直嗶聲來提醒擁有者。

2. 實驗目標：讓使用者利用 Lua 程式範例練習，同時讀取環境的溫度和濕度以及土壤的濕度值，再進行進階的判斷使蜂鳴器發出聲響提醒使用者。

3. 實驗步驟：此實驗將分兩個階段進行，

   a. 使用 Fritzing 實際產生麵包板電路圖、電路概要圖以及 PCB 印刷板電路圖，以及實際的麵包板接線圖。

   b. 加入蜂鳴器，調整 Lua 程式，當快缺水時，溫度過高或是過低時，濕度過高或是過低時，自動一直嗶聲來提醒擁有者。

a. 使用 Fritzing 實際產生麵包板電路圖、電路概要圖以及 PCB 印刷板電路圖，以及實際的麵包板接線圖

(1) 延續主題 C，搜尋"buzzer"，將蜂鳴器拉到麵包板上，如圖 7-29。

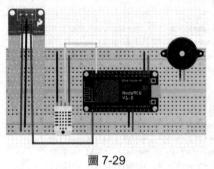

圖 7-29

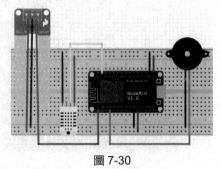

圖 7-30

(2) 將蜂鳴器的正極接腳接到 NodeMCU 上的 D12(SD3)數位接孔，負極接腳接到麵包板上的接地接腳，如圖 7-30。

(3) 點選上方的"概要圖"，再將元件以及線路擺放至較美觀的位置，並將線路佈線完成，如圖 7-31。

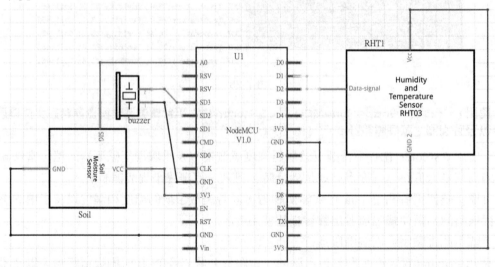

圖 7-31

(4) 點選上方的"PCB"，將元件以及線路擺放至較美觀的位置，並將線路佈線完成，如圖 7-32。

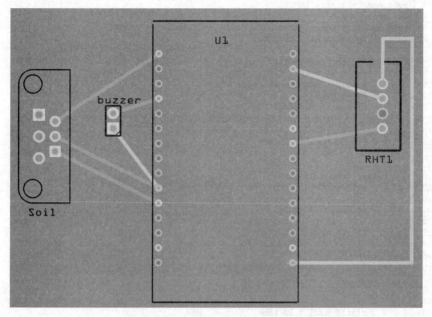

圖 7-32

(5) 接續主題 B 的實際麵包板接線，再將蜂鳴器的"+"接到 NodeMCU 上的 D12(SD3)接腳，GND 接到 NodeMCU 上的 GND 接腳，如圖 7-33。

圖 7-33

b. 加入蜂鳴器，調整 Lua 程式，當快缺水時，溫度過高或是過低時，濕度過高或是過低時，自動一直嗶聲來提醒擁有者

(1) 點擊 ESPlorer.bat 工具，進入開發環境 IDE。在左方的程式撰寫區撰寫 Lua 程式，如圖 7-34。當土壤濕度過高或過低時，利用 detect_temp_humi function 來偵測溫度過高或過低以及濕度過高或過低時的情況，再判斷是否讓蜂鳴器發出聲響。

```
1    pin = 1
2    soilpin = 0
3    gpio.mode(pin,gpio.INPUT)
4    buzzerpin = 12
5    gpio.mode(buzzerpin,gpio.OUTPUT)
6    gpio.write(buzzerpin,gpio.LOW)
7
8    function detect_temp_humi()
9        if temp > 30 then
10           gpio.write(buzzerpin,gpio.HIGH)
11           print("The temperature is too high!")
12           if humi > 80 then
13               --gpio.write(buzzerpin,gpio.HIGH)
14               print("The humidity is too high!")
15           elseif humi < 20 then
16               print("The humidity is too low!")
17           end
18       elseif temp < 24 then
19           gpio.write(buzzerpin,gpio.HIGH)
20           print("The temperature is too low!")
21           if humi > 80 then
22               print("The humidity is too high!")
23           elseif humi < 20 then
24               print("The humidity is too low!")
25           end
26       elseif temp>=24 and temp<=30 then
27           if humi > 80 then
28               gpio.write(buzzerpin,gpio.HIGH)
29               print("The humidity is too high!")
30           elseif humi < 20 then
31               gpio.write(buzzerpin,gpio.HIGH)
32               print("The humidity is too low!")
33           end
34       end
35   end
37   tmr.alarm(0,1000,1,function()
38       status, temp, humi, temp_dec, humi_dec = dht.read(pin)
39       soilstate = adc.read(soilpin)
40       print("DHT Temperature:"..temp.."; Humidity:"..humi.."; Soil:"..soilstate)
41       gpio.write(buzzerpin,gpio.LOW)
42       if status == dht.OK then
43           if soilstate > 700 then
44               gpio.write(buzzerpin,gpio.HIGH)
45               print("Soil humidity isn't enough!")
46               detect_temp_humi()
47           elseif soilstate <= 700 then
48               detect_temp_humi()
49           end
50       elseif status == dht.ERROR_CHECKSUM then
51           print("DHT Checksum error.")
52       elseif status == dht.ERROR_TIMEOUT then
53           print("DHT time out.")
54       end
55   end)
```

圖 7-34

(2) 實際麵包板接線方式如同步驟 a，詳細接線圖如圖 7-35。

(3) 到 ESPlorer IDE 平台點擊左下方"Save to ESP"按鈕，將程式存檔並寫入到 NodeMCU 後執行。

圖 7-35

4. 實驗成果：

a. 使用 Fritzing 實際產生麵包板電路圖、電路概要圖以及 PCB 印刷板電路圖，以及實際的麵包板接線圖

(1) 麵包板電路圖如圖 7-36。

(2) 電路概要圖如圖 7-37。

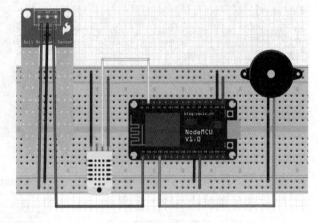

圖 7-36

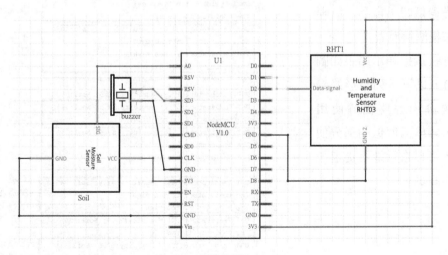

圖 7-37

93

(3) PCB 印刷板電路圖如圖 7-38。

(4) 實際的麵包板接線圖如圖 7-39。

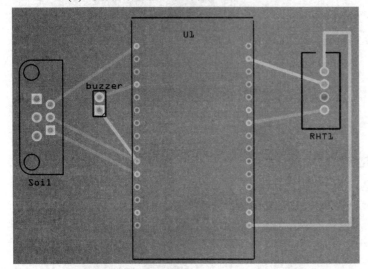

圖 7-38

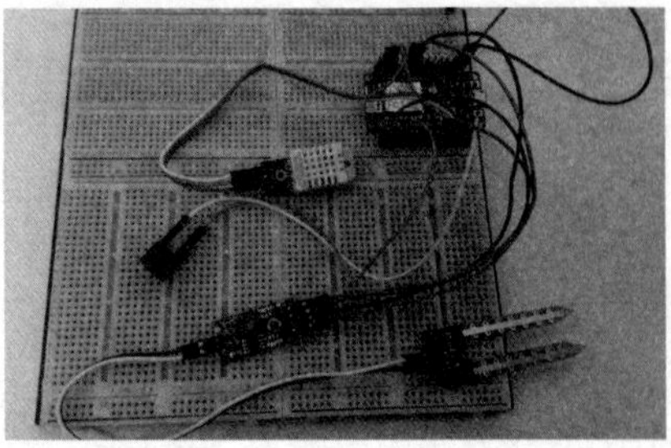

圖 7-39

b. 加入蜂鳴器，調整 Lua 程式，當快缺水時，溫度過高或是過低時，濕度過高或是過低時，自動一直嗶聲來提醒擁有者

部分終端機顯示結果如圖 7-40，以下所列情形蜂鳴器會發出聲響提醒使用者，關於溫度低的部分使用者可以自行進行更進一步的測試。

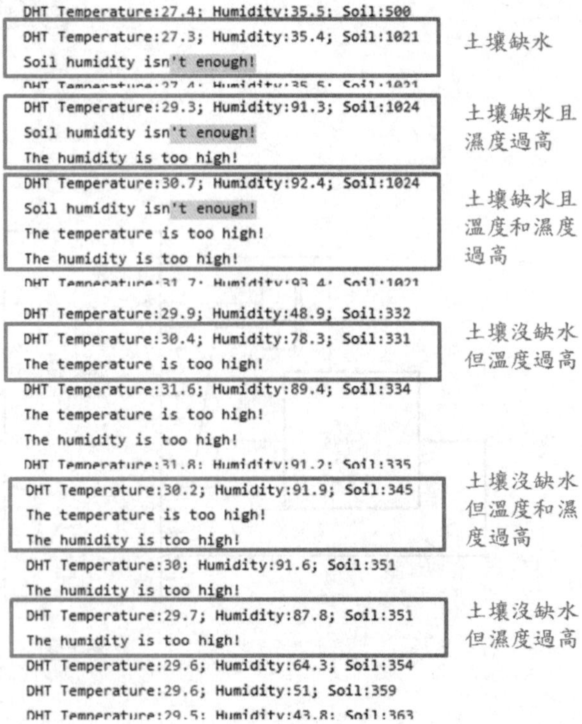

```
DHT Temperature:27.4; Humidity:35.5; Soil:500
DHT Temperature:27.3; Humidity:35.4; Soil:1021     土壤缺水
Soil humidity isn't enough!
DHT Temperature:27.4; Humidity:35.5; Soil:1021
DHT Temperature:29.3; Humidity:91.3; Soil:1024     土壤缺水且
Soil humidity isn't enough!                        濕度過高
The humidity is too high!
DHT Temperature:30.7; Humidity:92.4; Soil:1024     土壤缺水且
Soil humidity isn't enough!                        溫度和濕度
The temperature is too high!                       過高
The humidity is too high!
DHT Temperature:31.7; Humidity:93.4; Soil:1021
DHT Temperature:29.9; Humidity:48.9; Soil:332      土壤沒缺水
DHT Temperature:30.4; Humidity:78.3; Soil:331      但溫度過高
The temperature is too high!
DHT Temperature:31.6; Humidity:89.4; Soil:334
The temperature is too high!
The humidity is too high!
DHT Temperature:31.8; Humidity:91.2; Soil:335
DHT Temperature:30.2; Humidity:91.9; Soil:345      土壤沒缺水
The temperature is too high!                       但溫度和濕
The humidity is too high!                          度過高
DHT Temperature:30; Humidity:91.6; Soil:351
The humidity is too high!
DHT Temperature:29.7; Humidity:87.8; Soil:351      土壤沒缺水
The humidity is too high!                          但濕度過高
DHT Temperature:29.6; Humidity:64.3; Soil:354
DHT Temperature:29.6; Humidity:51; Soil:359
DHT Temperature:29.5; Humidity:43.8; Soil:363
```

圖 7-40

主題 A：使用 Fritzing IDE 平台做繼電器模組電路設計

1. 題目：使用 Fritzing 工具畫出 NodeMCU 與繼電器模組連接電路。

2. 實驗目標：使用繼電器模組範例練習 Fritzing 的電路設計，實際產生麵包板電路圖、電路概要圖以及 PCB 印刷板電路圖。

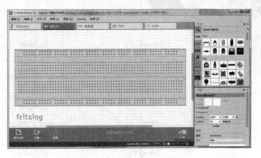

圖 8-1

3. 實驗步驟：此實驗將分三個階段進行，

   a. 產生麵包板電路圖。

   b. 產生電路概要圖。

   c. 產生 PCB 印刷板電路圖。

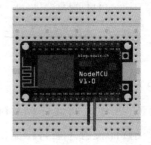

圖 8-2

a. 產生麵包板電路圖

   (1) 開啓 Fritzing 工具，點選"麵包板"，如圖 8-1。

   (2) 在右方搜尋窗格輸入"NodeMCU"，將 NodeMCU 拖曳至麵包板上並完成接線，如圖 8-2。

   (3) 搜尋元件"Relay"，將繼電器拖曳至麵包板上，如圖 8-3。

   (4) 將繼電器的 power 以及 ground 接腳分別接到麵包板上的電源和接地接孔，signal 接腳接到 NodeMCU 上的 D1 數位接孔，如圖 8-4。

圖 8-3

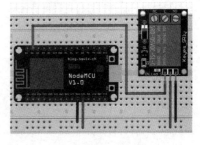

圖 8-4

b. 產生電路概要圖

　　點選上方的"概要圖"，將元件以及線路擺放至較美觀的位置，並將線路佈線完成，如圖 8-5。

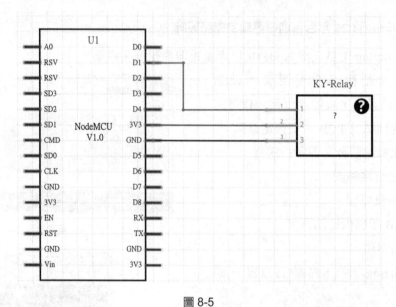

圖 8-5

c. 產生 PCB 印刷板電路圖

　　點選上方的"PCB"，將元件以及線路擺放至較美觀的位置，並將線路佈線完成，如圖 8-6。

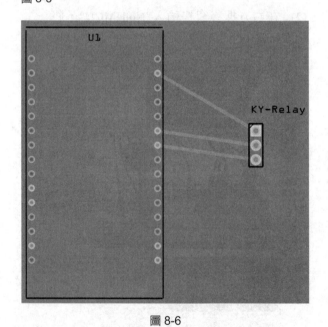

圖 8-6

4. 實驗成果：

    (1) 麵包板電路如圖 8-7。

    (2) 電路概要圖如圖 8-8。

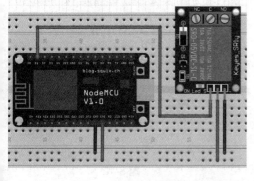

圖 8-7

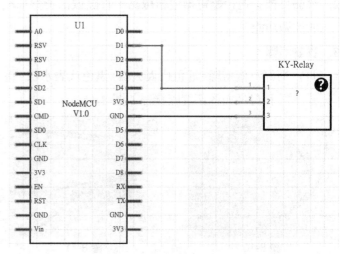

圖 8-8

    (3) PCB 印刷板電路圖如圖 8-9。

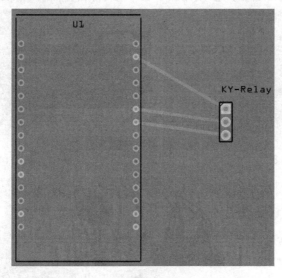

圖 8-9

主題 B：使用麵包板完成智慧插座

1. 題目：實際利用麵包板將繼電器模組進行線路連接完成一個智慧插座：

    (1) 剪三條線 (兩紅一黑)。

    (2) 將這些線連接公和母插頭。

2. 實驗目標：進行繼電器連結線路，實際做出智慧插座，以便之後利用程式控制智慧插座上電器的開關功能。

3. 實驗步驟：

    (1) 剪好三條電線，紅色代表火線，黑色代表水線。將水線兩端分別連接公和母插頭，如圖 8-10。公和母插頭內部實際接線如圖 8-11。

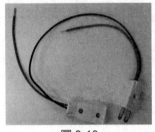
圖 8-10

圖 8-11

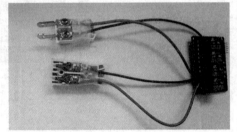
圖 8-12

    (2) 將兩條連接公和母接頭的火線連接到繼電器的 com 和 NO，如圖 8-12。實際將火線接到繼電器時，可以先將螺絲鬆開插入如圖 8-13；接著再將螺絲鎖緊如圖 8-14。

圖 8-13

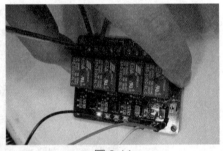
圖 8-14

    (3) 將繼電器模組的 "VCC" 接到 NodeMCU 上的 3V3 接孔，"GND" 接到 NodeMCU 上的 GND 接孔，"IN3" 接到 NodeMCU 上的 D1 數位接孔，如圖 8-15。這裡選擇"IN3"是因為使用的是繼電器模組的第三個。

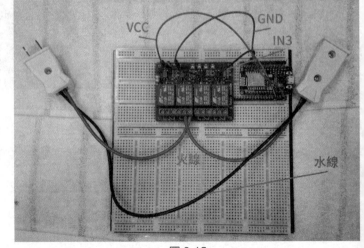
圖 8-15

主題 C：使用 ESPlorer IDE 平台實作繼電器模組程式設計

1. 題目：使用 ESPlorer IDE 平台撰寫 Lua 程式控制繼電器去開啟或關閉 AC 燈光，並觀察以及實驗下述兩點：

    (1) 先不接電路，測看看是否能夠開啟與關閉繼電器，會有"洽"的聲響。

    (2) 成功後，再接上電器測試。

2. 實驗目標：讓使用者利用 Lua 程式範例練習，如何控制繼電器開啟和關閉檯燈。

3. 實驗步驟：

    (1) 點擊 ESPlorer.bat 工具，進入開發環境 IDE。在左方的程式撰寫區撰寫 Lua 程式，如圖 8-16。

```
1  relayPin = 1
2  gpio.mode(relayPin, gpio.OUTPUT)
3  -- connect to NC
4  gpio.write(relayPin, gpio.HIGH)
5  -- connect to NO
6  gpio.write(relayPin, gpio.LOW)
```

圖 8-16

    (2) 將繼電器模組的"VCC"接到 NodeMCU 上的 3V3 接孔，"GND"接到 NodeMCU 上的 GND 接孔，"IN1"接到 NodeMCU 上的 D1 數位接孔，如圖 8-17。

圖 8-17

    (3) 到 ESPlorer IDE 平台點擊左下方"Save to ESP"按鈕，將程式存檔並寫入到 NodeMCU 後執行。

4. 實驗成果：

(1) 先不接電器，點選程式中設置高電位或低電位的單行程式後，點選"Line"，便能夠開啓與關閉繼電器，同時會有"洽"的聲響，如圖 8-18。

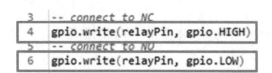 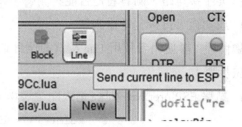

圖 8-18

(2) 確定能夠開啓與關閉繼電器成功後，再接上電器測試，如圖 8-19。

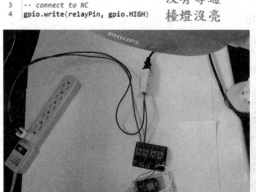 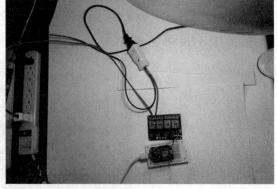

圖 8-19

主題 D：使用 Fritzing、Snap4NodeMCU 與 ESPlorer IDE 平台實作繼電器模組結合壓力感測器、超音波感測器或是移動感測器之程式設計

1. 題目：加入壓力感測器、超音波感測器或是移動感測器，使用 ESPlorer IDE 平台調整 Lua 程式，當偵測到有人的時候，自動點亮燈光。

2. 實驗目標：讓使用者利用 Lua 程式範例練習，利用壓力感測器、超音波感測器或是移動感測器偵測到有人，再利用繼電器控制檯燈的開或關。

3. 實驗步驟：此實驗將分兩個階段進行，

   a. 使用 Fritzing 實際產生麵包板電路圖、電路概要圖以及 PCB 印刷板電路圖，以及實際的麵包板接線圖。

b. 加入壓力感測器、超音波感測器或是移動感測器，調整 Lua 程式，當偵測到有人的時候，自動點亮燈光。

a. 使用 Fritzing 實際產生麵包板電路圖、電路概要圖以及 PCB 印刷板電路圖，以及實際的麵包板接線圖

(1) 延續主題 C，搜尋"PIR"，將移動感測器拉到麵包板上，如圖 8-20。

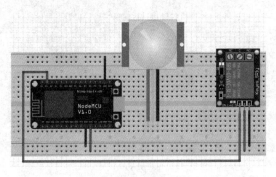
圖 8-20

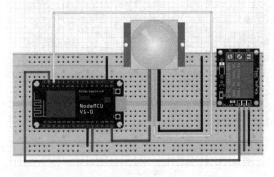
圖 8-21

(2) 將移動感測器的黃色接腳接到 NodeMCU 上的 D2 數位接孔，黑色接腳接到麵包板上的 GND，紅色接腳接到 NodeMCU 上的 Vin 接孔，如圖 8-21。

(3) 點選上方的"概要圖"，將元件以及線路擺放至較美觀的位置，並將線路佈線完成，如圖 8-22。

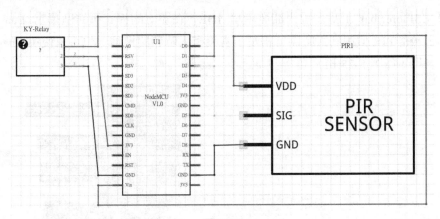
圖 8-22

(4) 點選上方的"PCB"，將元件以及線路擺放至較美觀的位置，並將線路佈線完成，如圖 8-23。

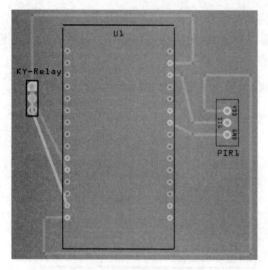

圖 8-23

圖 8-24

(5) 接續主題 C 的實際麵包板接線，再將移動感測器的"VCC"接到 NodeMCU 上的 Vin 接腳，"GND"接到 NodeMCU 上的 GND 接腳，"OUT"接到 NodeMCU 上的 D2 接腳，如圖 8-24。

b. 加入壓力感測器、超音波感測器或是移動感測器，調整 Lua 程式，當偵測到有人的時候，自動點亮燈光

(1) 點擊 ESPlorer.bat 工具，進入開發環境 IDE。在左方的程式撰寫區撰寫 Lua 程式，如圖 8-25(a)。當 PIR 偵測到人時開啟檯燈，當沒有人時關閉檯燈。也可以使用 Snap4NodeMCU 加以實現，如圖 8-25(b)。

```
1   relay=8                                    -- set relay pin to pin 8
2   gpio.mode(relay,gpio.OUTPUT)               -- set pin 8 as output pin
3   PIR=2                                       -- set PIR pinto pin 2
4   gpio.mode(PIR,gpio.INPUT)                   -- set pin 2 as input pin
5   tmr.alarm(0,500,tmr.ALARM_AUTO,function()   -- repeat every 0.5 second
6     state=gpio.read(PIR)                      -- read PIR (digital) pin
7     if (state == 1) then                      -- PIR detects someone is there
8       print("state=",state,"Light open")
9       gpio.write(relay,gpio.HIGH)             -- turn on the Relay (light)
10    else
11      print("state=",state,"Light close")
12      gpio.write(relay,gpio.LOW)              -- turn off the relay
13    end
14  end)
```

圖 8-25(a)

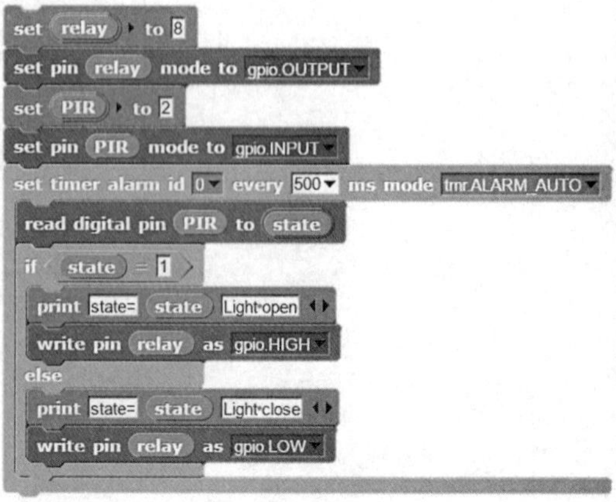

圖 8-25(b)

(2) 實際麵包板接線方式如同步驟 a，詳細接線圖如圖 8-26。

圖 8-26

(3) 到 ESPlorer IDE 平台點擊左下方"Save to ESP"按鈕，將程式存檔並寫入到 NodeMCU 後執行。

4. 實驗成果：

a. 使用 Fritzing 實際產生麵包板電路圖、電路概要圖以及 PCB 印刷板電路圖，以及實際的麵包板接線圖

  (1) 麵包板電路圖如圖 8-27。

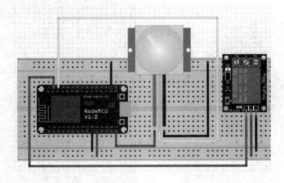

圖 8-27

(2) 電路概要圖如圖 8-28。

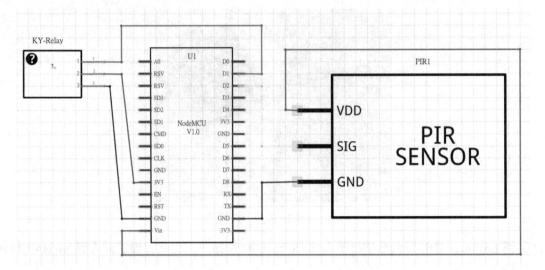

圖 8-28

(3) PCB 印刷板電路圖如圖 8-29。

(4) 實際的麵包板接線圖如圖 8-30。

圖 8-29

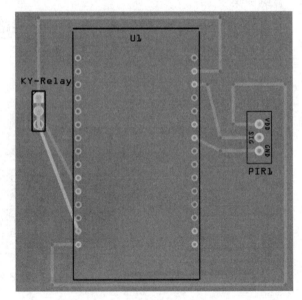

圖 8-30

b. 加入壓力感測器、超音波感測器或是移動感測器，調整 Lua 程式，當偵測到有人的時候，自動
   點亮燈光
   終端機顯示結果如圖 8-31，當偵測到有人時燈會打開；沒有偵測到人時燈會熄滅，實際檯燈明
   暗使用者可以自行進行更進一步的測試。

```
state=0 Light close      state=0 Light close
state=0 Light close      state=0 Light close
state=0 Light close      state=0 Light close
state=0 Light close      state=1 Light open
state=0 Light close      state=1 Light open
state=0 Light close      state=1 Light open
state=0 Light close      state=1 Light open
state=0 Light close      state=1 Light open
state=0 Light close      state=1 Light open
state=0 Light close      state=1 Light open
state=0 Light close      state=1 Light open
state=0 Light close      state=1 Light open
```

圖 8-31

主題 A：使用 Fritzing IDE 平台做壓力感測器電路設計

1. 題目：使用 Fritzing 工具畫出 NodeMCU 與壓力感測器連接電路。

2. 實驗目標：使用壓力感測器範例練習 Fritzing 的電路設計，實際產生麵包板電路圖、電路概要圖以及 PCB 印刷板電路圖。

3. 實驗步驟：此實驗將分三個階段進行，

   a. 產生麵包板電路圖。

   b. 產生電路概要圖。

   c. 產生 PCB 印刷板電路圖。

a. 產生麵包板電路圖

   (1) 開啓 Fritzing 工具，點選"麵包板"，如圖 9-1。

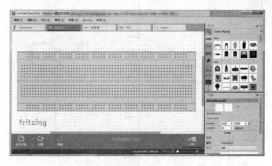

圖 9-1

   (2) 在右方搜尋窗格輸入"NodeMCU"，將 NodeMCU 拖曳至麵包板上並完成接線，如圖 9-2。

   (3) 搜尋元件"FSR"和電阻元件"resistor"，將壓力感測器和電阻拖曳至麵包板上，此時電阻的一邊接腳要和壓力感測器的一端串接在一起，如圖 9-3。

圖 9-2

   (4) 將壓力感測器和電阻串接的接腳接到 NodeMCU 上的 A0 數位接孔，壓力感測器的另一端接腳接到麵包板上的接地接孔。而電阻的另一端接腳接到麵包板上的電源接孔，如圖 9-4。

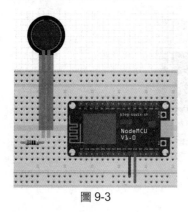

圖 9-3

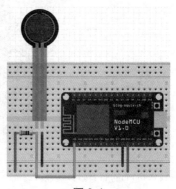

圖 9-4

b. 產生電路概要圖

點選上方的"概要圖"，再將元件以及線路擺放至較美觀的位置，並將線路佈線完成，如圖 9-5。

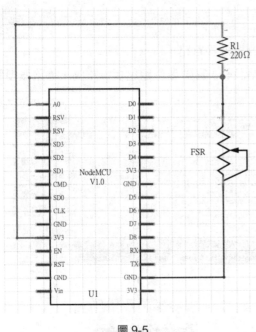

圖 9-5

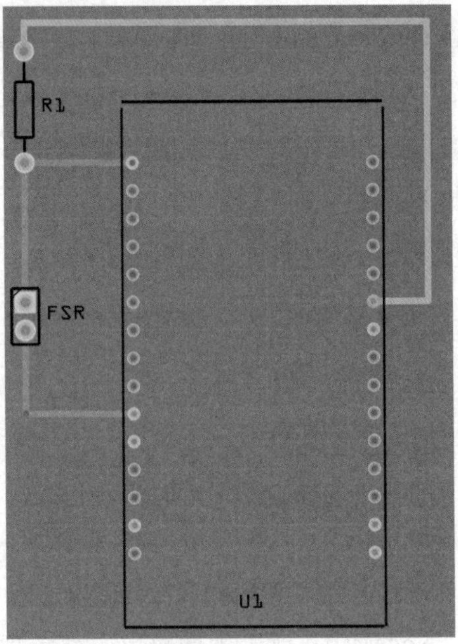

圖 9-6

c. 產生 PCB 印刷板電路圖

點選上方的"PCB"，將元件以及線路擺放至較美觀的位置，並將線路佈線完成，如圖 9-6。

4. 實驗成果：

(1) 麵包板電路如圖 9-7。

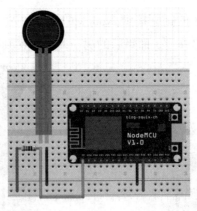

圖 9-7

(2) 電路概要圖如圖 9-8。

(3) PCB 印刷板電路圖如圖 9-9。

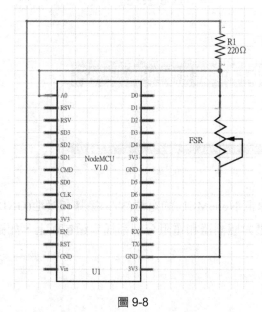

圖 9-8

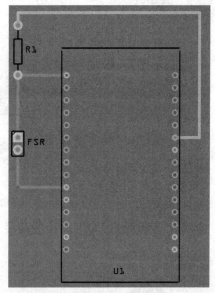

圖 9-9

主題 B：使用 Snap4NodeMCU 與 ESPlorer IDE 平台做壓力感測器程式設計

1. 題目：實際利用麵包板將 NodeMCU 以及壓力感測器進行線路連接，並使用 ESPlorer IDE 平台撰寫讀取壓力感測器的值，並觀察以及實驗下述兩點：

   (1) 輕輕壓下壓力感測器，觀察終端機顯示的變化。

   (2) 用力壓下壓力感測器，觀察終端機顯示的變化。

2. 實驗目標：使用讀取壓力感測器之範例程式練習 Lua 的 C 設計，並實際觀察麵包板接線完成後，程式進行時 ESPlorer 終端機顯示畫面，讓使用者實際了解當施壓的壓力不同時，實際壓力讀取值的變化情形。

3. 實驗步驟：

   (1) 點擊 ESPlorer.bat 工具，進入開發環境 IDE。在左方的程式撰寫區撰寫 Lua 程式，每 0.2 秒呼叫函數傳回壓力的讀取值，如圖 9-10(a)。也可以使用 Snap4NodeMCU 加以實現，如圖 9-10(b)。

```
1  ⊟tmr.alarm(0,200,tmr.ALARM_AUTO,function()    -- repeat every 0.2 second
2      force=adc.read(0)                          -- read analog value
3      print("force=",force)                      -- debug force
4   └end)
```

圖 9-10(a)

```
set timer alarm id 0 ▾ every 200 ▾ ms mode tmr.ALARM_AUTO ▾
    read analog pin to  force
    print  force=   force  ◀▶
```

圖 9-10(b)

(2) 將 NodeMCU 插入到麵包板上，再利用杜邦線把 NodeMCU 的電源(3V3)接腳和接地(GND)
    接腳接到麵包板上的電源和接地接孔。再將壓力感測器和電阻分別都插到麵包板上，如圖
    9-11。

圖 9-11

圖 9-12

(3) 將壓力感測器和電阻串接的接腳接到 NodeMCU 上的 A0 類比接孔，壓力感測器的另一端
    接腳接到 NodeMCU 上的 GND 接孔。而電阻的另一端接腳接到 NodeMCU 上的 3V3 接孔，
    如圖 9-12。

(4) 到 ESPlorer IDE 平台點擊左下方"Save to ESP"按鈕，將程式存檔並寫入到 NodeMCU 後執
    行。

4. 實驗成果：

　(1) 輕輕壓下壓力感測
　　　器，終端機顯示的變
　　　化如圖 9-13。虛線下
　　　方為開始輕壓。

　(2) 用力壓下壓力感測
　　　器，終端機顯示的變
　　　化如圖 9-14。虛線下
　　　方為開始重壓。

Force = 1024	Force = 1024
Force = 1024	Force = 1024
Force = 1024	Force = 1024
Force = 1024	Force = 1024
Force = 1024	Force = 1024
Force = 1024	Force = 1024
Force = 1023	Force = 884
Force = 1022	Force = 828
Force = 1020	Force = 792
Force = 1017	Force = 788
Force = 1009	Force = 785
Force = 1001	Force = 786
Force = 1002	Force = 776
Force = 993	Force = 775
Force = 993	Force = 775
Force = 992	Force = 773
Force = 991	Force = 768
Force = 989	Force = 767
Force = 988	Force = 768
Force = 989	Force = 767
Force = 989	
Force = 988	

圖 9-13　　　　　　　　　　　　圖 9-14

主題 C：使用 Fritzing、Snap4NodeMCU 與 ESPlorer IDE 平台做壓力感測器結合 LED 燈或蜂鳴器之設計

1. 題目：使用 ESPlorer IDE 平台撰寫讀取壓力感測器的值，進行判斷後開啟 LED 燈或蜂鳴器，並改變燈亮明暗或是聲音大小。

2. 實驗目標：讓使用者利用 Lua 程式範例練習，藉由讀取壓力值判斷壓力大小而對 LED 或蜂鳴器做進一步的控制。

3. 實驗步驟：此實驗將分兩個階段進行，

　　a. 使用 Fritzing 實際產生壓力感測器結合 LED 燈的麵包板電路圖、電路概要圖以及 PCB 印刷板電路圖，以及實際的麵包板接線圖。

　　b. 調整 Lua 程式，當壓力感測器偵測到有人的時，自動啟動 LED 燈或蜂鳴器。

　　c. 調整 Lua 程式，利用壓力感測器調整 LED 燈亮度或蜂鳴器大小聲。

a. 使用 Fritzing 實際產生壓力感測器結合 LED 燈的麵包板電路圖、電路概要圖以及 PCB 印刷板電路圖，以及實際的麵包板接線圖

　(1) 延續主題 B，搜尋"LED"和"resistor"，將 LED 燈和電阻拉到麵包板上，如圖 9-15。

111

(2) 將 LED 的長邊接腳接到
NodeMCU 上的 D2 數位接
孔，短邊接腳和電阻相接，
而電阻的另一端接腳接到
麵包板上的 GND，如圖
9-16。

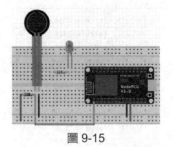

圖 9-15

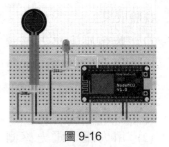

圖 9-16

(3) 點選上方的"概要圖"，將元件以及線路擺放至較美觀的位置，並將線路佈線完成，如圖 9-17。

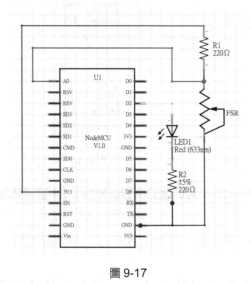

圖 9-17

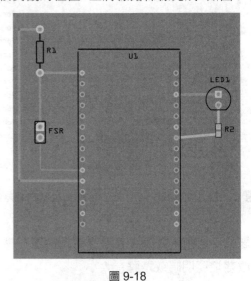

圖 9-18

(4) 點選上方的"PCB"，將元件以及線路擺放至較美觀的位置，並將線路佈線完成，如圖 9-18。

(5) 接續主題 B 的實際麵包板接線，將 LED
的長邊接腳接到 NodeMCU 上的 D2 數位
接孔，短邊接腳和電阻相接，而電阻的
另一端接腳接到麵包板上的 GND，如圖
9-19。

圖 9-19

b. 調整 Lua 程式，當壓力感測器偵測到有人的時，自動啓動 LED 燈或蜂鳴器

(1) 點擊 ESPlorer.bat 工具，進入開發環境 IDE。在左方的程式撰寫區撰寫 Lua 程式，如圖 9-20(a)。也可以使用 Snap4NodeMCU 加以實現，如圖 9-20(b)。

```
1  led=4                                          -- set led to pin 4
2  gpio.mode(led,gpio.OUTPUT)                      -- set pin 4 as output pin
3  tmr.alarm(0,200,tmr.ALARM_AUTO,function()       -- repeat every 0.2 second
4    force=adc.read(0)                             -- read analog value
5    if (force < 800) then                         -- if there is a person detecetd
6      print("force=",force,"Light ON")            -- turn on the light
7      gpio.write(led,gpio.HIGH)
8    else
9      print("force=",force,"Light OFF")           -- turn off the light
10     gpio.write(led,gpio.LOW)
11   end
12 end)
```

圖 9-20(a)

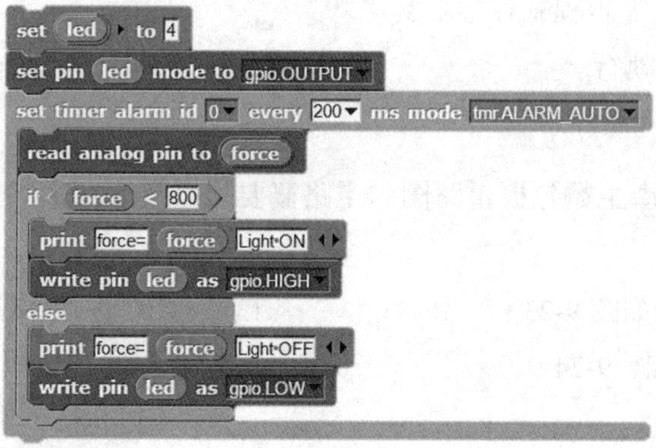

圖 9-20(b)

(2) 實際麵包板接線方式如同步驟 a，詳細接線圖如圖 9-21。

圖 9-21

(3) 到 ESPlorer IDE 平台點擊左下方"Save to ESP"按鈕，將程式存檔並寫入到 NodeMCU 後執行。

c. 調整 Lua 程式，利用壓力感測器調整 LED
   燈亮度或蜂鳴器大小聲

   (1) 修改步驟 b 的程式碼，將 force < 950
       設定為偵測到有人。接著判斷如果壓
       力不變或持續增加時將 LED 燈逐漸
       打亮；若是有人但是壓力變小的話，
       則 LED 燈維持一定亮度；若是沒有人
       時，則將 LED 燈漸漸熄滅。程式碼範
       例如圖 9-22。

   (2) 到 ESPlorer IDE 平台點擊左下方
       "Save to ESP"按鈕，將程式存檔並寫
       入到 NodeMCU 後執行。

```lua
1  pin = 0
2  ledpin = 2
3  width = 0
4  clock = 1000
5  force=1024
6  pwm.setup(ledpin,clock,width)
7  pwm.start(ledpin)
8  tmr.alarm(0, 200, 1, function()
9      force_lasttime=force
10     force = adc.read(pin)
11     force_diff=force-force_lasttime
12     if (force < 950) and (force_diff<=0) then
13         if (width < 1000) then
14             width = width + 50
15         end
16     elseif (force < 950) and (force_diff>0) then
17             width = width
18     else
19         if (width>0) then
20             width=width-100
21         end
22     end
23     pwm.setduty(ledpin,width)
24     print("force value=" ..force..",width = " ..width)
25 end)
```

圖 9-22

4. 實驗成果：

   a. 使用 Fritzing 實際產生麵包板電路圖、電路概要圖以及 PCB 印刷板電路圖，以及實際的麵包板接線圖

      (1) 麵包板電路圖如圖 9-23。   (3) PCB 印刷板電路圖如圖 9-25。

      (2) 電路概要圖如圖 9-24。   (4) 實際的麵包板接線圖如圖 9-26。

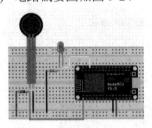

圖 9-23

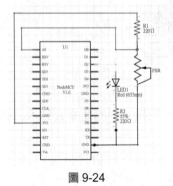

圖 9-24

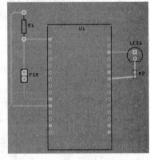

圖 9-25

圖 9-26

b. 調整 Lua 程式，當壓力感測器偵測到有人的時，自動啟動 LED 燈或蜂鳴器

終端機顯示結果如圖 9-27。虛線上方是當壓力感測器沒有偵測到人時的壓力讀取值以及 LED 燈是暗的，虛線下方則是輕壓壓力感測器模擬有人時的壓力讀取值以及 LED 燈是亮的情形。

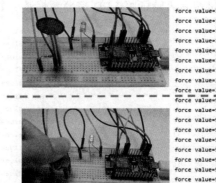

```
force value=1022,Light OFF
force value=1024,Light OFF
force value=1022,Light OFF
force value=1022,Light OFF
force value=1022,Light OFF
force value=1022,Light OFF
force value=1022,Light OFF
force value=1021,Light OFF
- - - - - - - - - - - - - -
force value=981,Light ON
force value=939,Light ON
force value=934,Light ON
force value=932,Light ON
force value=937,Light ON
force value=938,Light ON
force value=939,Light ON
force value=931,Light ON
force value=929,Light ON
force value=927,Light ON
force value=927,Light ON
force value=923,Light ON
force value=926,Light ON
```

圖 9-27

c. 調整 Lua 程式，利用壓力感測器調整 LED 燈亮度或蜂鳴器大小聲

(1) 偵測到有人而且壓力不變或持續增加時，會將 LED 燈逐漸打亮，終端機顯示結果如圖 9-28。LED 燈逐漸打亮情形請實際操作練習觀察。

```
force value=1024,width = 0
force value=1024,width = 0
force value=878,width = 50
force value=796,width = 100
force value=775,width = 150
force value=763,width = 200
force value=754,width = 250
force value=742,width = 300
force value=737,width = 350
force value=737,width = 400
force value=737,width = 450
force value=732,width = 500
force value=731,width = 550
force value=729,width = 600
force value=729,width = 650
force value=728,width = 700
force value=726,width = 750
force value=724,width = 800
force value=724,width = 850
force value=721,width = 900
force value=722,width = 900
force value=722,width = 950
force value=721,width = 1000
```

圖 9-28

```
force value=721,width = 900
force value=722,width = 900
force value=722,width = 950
force value=721,width = 1000
force value=721,width = 1000
force value=720,width = 1000
force value=720,width = 1000
force value=720,width = 1000
force value=720,width = 1000
force value=724,width = 1000
force value=753,width = 1000
force value=785,width = 1000
force value=875,width = 1000
force value=958,width = 900
force value=1005,width = 800
```

圖 9-30

(2) 若是有人但是壓力變小的話，則 LED 燈維持一定亮度，如圖 9-30 紅框處。

115

(3) 若是沒有人時則將 LED 燈漸漸熄滅。如圖 9-31。

```
force value=785,width = 1000
force value=875,width = 1000
force value=958,width = 900
force value=1005,width = 800
force value=1024,width = 700
force value=1023,width = 600
force value=1023,width = 500
force value=1024,width = 400
force value=1024,width = 300
force value=1024,width = 200
force value=1024,width = 100
force value=1024,width = 0
force value=1024,width = 0
```

圖 9-31

主題 A：使用 Fritzing IDE 平台做聲音感測模組電路設計

1. 題目：使用 Fritzing 工具畫出 NodeMCU 與聲音感測模組連接電路。

2. 實驗目標：使用聲音感測模組範例練習 Fritzing 的電路設計，實際產生麵包板電路圖、電路概要圖以及 PCB 印刷板電路圖。

3. 實驗步驟：此實驗將分三個階段進行，

   a. 產生麵包板電路圖。

   b. 產生電路概要圖。

   c. 產生 PCB 印刷板電路圖。

a. 產生麵包板電路圖

   (1) 開啟 Fritzing 工具，點選"麵包板"，如圖 10-1。

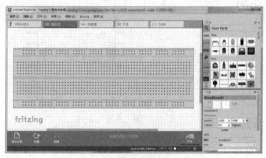

圖 10-1

   (2) 在右方搜尋窗格輸入 "NodeMCU"，將 NodeMCU 拖曳至麵包板上並完成接線，如圖 10-2。

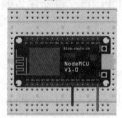

圖 10-2

   (3) 由於 Fritzing 工具目前沒有提供 Keyes 的高感度聲音感測模組，因此使用"Breakout Board for Electret Microphone"來進行線路圖的實作。搜尋元件"microphone"，將聲音感測模組拖曳至麵包板上，如圖 10-3。

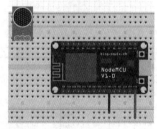

圖 10-3

   (4) 將聲音感測模組的 VCC 以及 GND 接腳分別接到麵包板上的電源和接地接孔，AUD 接腳接到 NodeMCU 上的 A0 接孔，如圖 10-4。

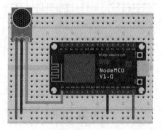

圖 10-4

b. 產生電路概要圖

點選上方的"概要圖"，將元件以及線路擺放至較美觀的位置，並將線路佈線完成，如圖 10-5。

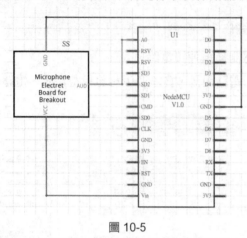

圖 10-5

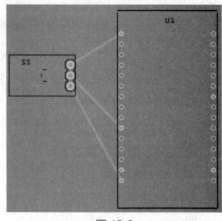

圖 10-6

c. 產生 PCB 印刷板電路圖

點選上方的"PCB"，將元件以及線路擺放至較美觀的位置，並將線路佈線完成，如圖 10-6。

4. 實驗成果：

(1) 麵包板電路如圖 10-7。

(2) 電路概要圖如圖 10-8。

(3) PCB 印刷板電路圖如圖 10-9。

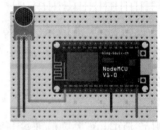

圖 10-7

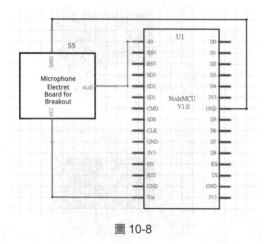

圖 10-8

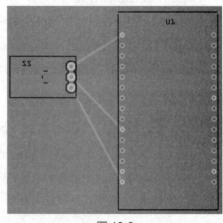

圖 10-9

主題 B：使用 Snap4NodeMCU 與 ESPlorer IDE 平台做聲音感測模組程式設計

1. 題目：實際利用麵包板將 NodeMCU 以及聲音感測模組進行線路連接，並使用 ESPlorer IDE 平台撰寫讀取聲音感測模組的值，並觀察以及實驗下述三點：

    (1) 調整靈敏度，觀察 L2 的 LED 燈以及終端機顯示的變化。

    (2) 輕拍聲音感測模組，觀察終端機顯示的變化。

    (3) 播放音樂，觀察終端機顯示的變化。

2. 實驗目標：使用讀取聲音感測模組之範例程式練習 Lua 的程式設計，並實際觀察麵包板接線完成後，程式進行時 ESPlorer 終端機顯示畫面，讓使用者實際了解聲音感測模組接收聲音時的變化情形。

3. 實驗步驟：

    (1) 點擊 ESPlorer.bat 工具，進入開發環境 IDE。在左方的程式撰寫區撰寫 Lua 程式，每 0.05 秒呼叫函數傳回聲音感測模組的讀取值，如圖 10-10(a)。也可以使用 Snap4NodeMCU 加以實現，如圖 10-10(b)。

```
1  tmr.alarm(0,50,tmr.ALARM_AUTO,function()    -- repeat every 0.2 second
2     aValue=adc.read(0)                        -- read analog value
3     print("sound=",aValue)                    -- print out sound value
4  end)
```

圖 10-10(a)

圖 10-10(b)

(2) 將 NodeMCU 插入到麵包板上，再利用杜邦線把 NodeMCU 的電源(3V3)接腳和接地(GND)接腳接到麵包板上的電源和接地接孔。將聲音感測模組的"+"接到 NodeMCU 上的 Vin 接孔，"G"接到 NodeMCU 上的 GND 接孔，"A0"接到 NodeMCU 上的 A0 接孔，如圖 10-11。
註：電源若是接 3V3 的話，聲音感測模組的讀取值靈敏度較低。

圖 10-11

(3) 到 ESPlorer IDE 平台點擊左下方"Save to ESP"按鈕，將程式存檔並寫入到 NodeMCU 後執行。

4. 實驗成果：

(1) 順時針調整靈敏度時聲音讀取值逐漸變小；逆時針調整時聲音讀取值則逐漸變大，終端機顯示的變化如圖 10-13。一開始接上電源時，Power 指示燈會亮，如圖 10-12 左；當調整靈敏度聲音讀取值約為 520 時，L2 的指示燈也會亮起，如圖 10-12 右。

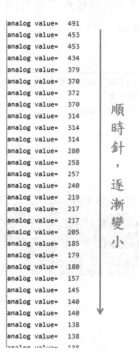

```
analog value=  491        analog value=  154
analog value=  453        analog value=  162
analog value=  453        analog value=  183
analog value=  434        analog value=  200
analog value=  379        analog value=  201
analog value=  370        analog value=  208
analog value=  372        analog value=  222
analog value=  370        analog value=  258
analog value=  314        analog value=  257
analog value=  314        analog value=  257
analog value=  314        analog value=  285
analog value=  280        analog value=  315
analog value=  258        analog value=  323
analog value=  257        analog value=  324
analog value=  240        analog value=  325
analog value=  219        analog value=  340
analog value=  217        analog value=  367
analog value=  217        analog value=  389
analog value=  205        analog value=  407
analog value=  185        analog value=  408
analog value=  179        analog value=  411
analog value=  180        analog value=  413
analog value=  157        analog value=  448
analog value=  145        analog value=  464
analog value=  140        analog value=  501
analog value=  140        analog value=  505
analog value=  138        analog value=  507
analog value=  138        analog value=  504
```

順時針，逐漸變小　　逆時針，逐漸變大

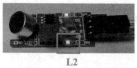

Power

L2

圖 10-12

圖 10-13

(2) 輕拍聲音感測模組，終端機顯示的變化如圖 10-14。以目前調整的靈敏度，在一般環境噪音下讀到的數位值介於 510 到 516 之間，當輕拍感測模組可達 562 或 564。

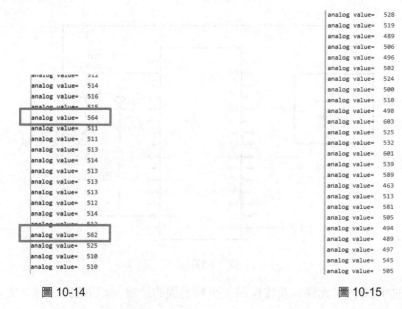

圖 10-14　　　　　　　　　　　　　　　　　　圖 10-15

(3) 播放音樂，終端機顯示的變化如圖 10-15。當播放音樂時數值的起伏比較大。

主題 C：使用 Fritzing、Snap4NodeMCU 與 ESPlorer IDE 平台做聲音感測模組結合 LED 軟硬體設計

1. 題目：使用 ESPlorer IDE 平台撰寫讀取聲音感測模組的值並動態呈現 LED。
2. 實驗目標：讓使用者利用 Lua 程式範例練習，藉由播放音樂時動態的讓 LED 燈閃爍。
3. 實驗步驟：此實驗將分兩個階段進行，
   a. 使用 Fritzing 實際產生麵包板電路圖、電路概要圖以及 PCB 印刷板電路圖，以及實際的麵包板接線圖。
   b. 加入數個 LED 燈，撰寫 Lua 程式當播放音樂時，可以動態呈現 LED 燈。

a. 使用 Fritzing 實際產生麵包板電路圖、電路概要圖以及 PCB 印刷板電路圖，以及實際的麵包板接線圖

   (1) 延續主題 B，將多個 LED 和電阻拉到麵包板上並接線，在此實作中以兩個 LED 為例，如圖 10-16。

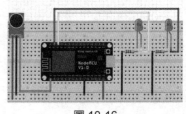

圖 10-16

(2) 點選上方的"概要圖"，將元件以及線路擺放至較美觀的位置，並將線路佈線完成，如圖 10-17。

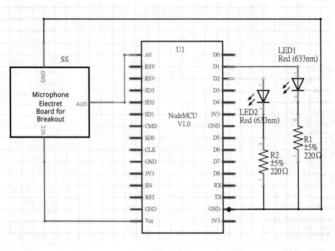

圖 10-17

(3) 點選上方的"PCB"，將元件以及線路擺放至較美觀的位置，並將線路佈線完成，如圖 10-18。

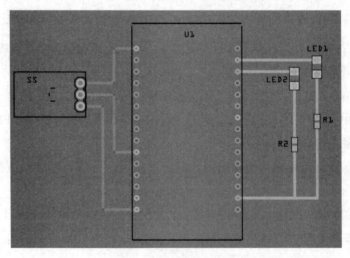

圖 10-18

(4) 接續主題 B 的實際麵包板接線,再將多
    個 LED 燈和電阻接到麵包板上。LED
    的長邊腳位分別接到 NodeMCU 上的
    D1、D2、D3 接腳,每個 LED 的短邊接
    腳和電阻一端相接,而電阻的另一端接
    腳則接到麵包板上的接地接孔,如圖
    10-19。

圖 10-19

b. 加入數個 LED 燈,撰寫 Lua 程式當播放音樂
   時,可以動態呈現 LED 燈

```
1   soundpin = 0
2   ledyellow=1
3   ledgreen=2
4   ledred=3
5   tmr.alarm(0,50,1,function()
6       sound=adc.read(soundpin)
7       print("analog value="..sound)
8       if sound >= 800 then
9           gpio.write(ledyellow, gpio.HIGH)
10      elseif sound < 800 and sound >= 790 then
11          gpio.write(ledred, gpio.HIGH)
12      elseif sound < 790 then
13          gpio.write(ledgreen, gpio.HIGH)
14      end
15      gpio.write(ledyellow, gpio.LOW)
16      gpio.write(ledgreen, gpio.LOW)
17      gpio.write(ledred, gpio.LOW)
18  end)
```

圖 10-20

   (1) 點擊 ESPlorer.bat 工具,進入開發環境
       IDE,在左方的程式撰寫區撰寫 Lua 程
       式,如圖 10-20。

   (2) 實際麵包板接線方式如同步驟 a,詳細
       接線圖如圖 10-21。

   (3) 到 ESPlorer IDE 平台點擊左下方"Save
       to ESP"按鈕,將程式存檔並寫入到
       NodeMCU 後執行。

圖 10-21

4. 實驗成果:
   a. 使用 Fritzing 實際產生麵包板電路圖、
      電路概要圖以及 PCB 印刷板電路圖,
      以及實際的麵包板接線圖
      (1) 麵包板電路圖如圖 10-22。

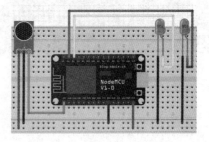

圖 10-22

(2) 電路概要圖如圖 10-23。

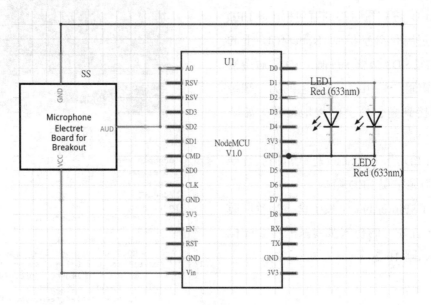

圖 10-23

(3) PCB 印刷板電路圖如圖 10-24。

(4) 實際的麵包板接線圖如圖 10-25。

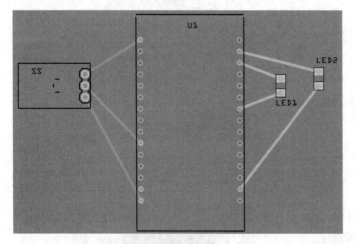

圖 10-24

圖 10-25

b. 加入數個 LED 燈，撰寫 Lua 程式當播放音
   樂時，可以動態呈現 LED 燈
   終端機顯示結果如圖 10-26。當聲音讀取值
   大於等於 800 亮黃燈，介於 790 到 800 之間
   亮紅燈，小於 790 則亮綠燈。

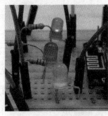

```
analog value=792
analog value=796
analog value=793
analog value=789    綠燈
analog value=791
analog value=798
analog value=799
analog value=791
analog value=791
analog value=793
analog value=793
analog value=791
analog value=792
analog value=795
analog value=791
analog value=791
analog value=803    黃燈
analog value=794
analog value=791
analog value=793
```

圖 10-26

主題Ａ：實做微波爐電路設計

1. 題目：實際利用麵包板將微處理器(MCU)、光感測器、矩陣鍵盤、LED 燈以及繼電器進行線路連接，實做微波爐電路。

2. 實驗目標：利用嵌入式系統的感測器(光感測器與矩陣鍵盤)與觸動器(LED 燈與繼電器)實現微波爐的雛形。

    (1) 光感測器擬偵測開門或關門。

    (2) 矩陣鍵盤感測器模擬微波爐的控制面板。

    (3) LED 燈模擬微波爐內部的燈光。

    (4) 繼電器模擬啓動微波加熱器(heater)與加熱轉盤(turner)。

3. 實驗步驟：

    (1) 光感測器擬繼電器模組電路設計請參考「單元三：光敏電阻以及類比數位轉換器原理與應用」，將光感測器控制端接到 NodeMCU 的 A0(第 0 類比腳位)，如圖 11-1。

```
1  pin=0
2  gpio.mode(pin, gpio.OUTPUT)
3  gpio.write(pin, gpio.LOW)
4  wifi.setmode(wifi.STATION)
5  wifi.sta.config("4Clab-2.4G", "socad3284")
6  tmr.alarm(0,1000,1, function()
7    print(wifi.sta.getip())
8    if wifi.sta.getip()~=nil then
9      gpio.write(pin, gpio.HIGH)
10     tmr.stop(0)
11   end
12 end)
```

圖 11-1

    (2) 矩陣鍵盤感測器電路設計請參考「單元五：矩陣鍵盤感測器原理與應用」，將矩陣鍵盤控制端接到 NodeMCU 的 D1~D8(第 1~8 數位腳位)，如圖 11-2。

圖 11-2

(3) 繼電器模組電路設計請參考「單元八：繼電器原理與應用」，將繼電器控制端接到 NodeMCU 的 D12(第 12 數位腳位)，如圖 11-3。

圖 11-3

圖 11-4

(4) LED 燈電路設計電路設計請參考「單元二：PWM 原理與應用」，將 LED 燈控制端接到 NodeMCU 的 D0(第 0 數位腳位)，如圖 11-4。

(5) 微波爐電路設計，如圖 11-5。

圖 11-5

主題 B：使用 Snap4NodeMCU 與 ESPlorer IDE 平台做微波爐程式設計

1. 題目：使用 Snap4NodeMCU 與 ESPlorer IDE 平台撰寫微波爐程式，並觀察以及實驗下述四點：：
   (1) 不蓋/蓋住光感測器，觀察終端機顯示的變化，並檢查看看 LED 燈是否會開啟與關閉。
   (2) 利用鍵盤輸入 A、B、C 和 D 按鍵，設定微波爐烹煮的強度。
   (3) 利用鍵盤輸入 0~9 按鍵，設定微波爐烹煮的微波的時間秒數。
   (4) 利用鍵盤輸入*按鍵，啟動微波爐繼電器烹煮設定的秒數。
2. 實驗目標：撰寫微波爐的程式，並實際觀察微波爐接線完成後，微波爐的各項功能是否正確。

3. 實驗步驟：

(1) 利用 ESPlorer IDE 平台實做出微波爐程式，如圖 11-6。

```
1    relayPin=12                          -- set relay pin
2    gpio.mode(relayPin,gpio.OUTPUT)      -- set relay pin as output pin
3    gpio.write(relayPin,gpio.LOW)        -- set relay pin as LOW
4    relayInSec,strength=0                -- set cooking second and strength
5    keys={{"1","2","3","A"},{"4","5","6","B"},{"7","8","9","C"},{"*","0","#","D"}}
6    for i=1,8 do
7       gpio.mode(i,gpio.OUTPUT)          -- set keypad pin as output
8       gpio.write(i,gpio.LOW)            -- initialize to LOW
9    end
10   for i=5,8 do
11      gpio.mode(i,gpio.INPUT)           -- set pin 5~8 as input pin
12   end
13   tmr.alarm(0,200,tmr.ALARM_AUTO,function()
14      for r=1,4 do                      -- keypad scan algorithm
15         gpio.write(r,gpio.HIGH)        -- set row pin to HIGH
16         for c=1,4 do
17            dValue=gpio.read((c + 4))    -- scan/read column pin
18            if (dValue == 1) then        -- if hit (key is pressed)
19            print("r=",r,"c=",c,"key=",keys[r][c])
20            gpio.write(r,gpio.LOW)       -- set the row pin to LOW immediately
21            keypadFunc(keys[r][c])       -- call key pressed function
22            end
23         end
24         gpio.write(r,gpio.LOW)         -- set row pin to LOW
25      end
26   end)
27   function keypadFunc(key)
28      print("sec=",relayInSec,"s=",strength,"k=",key)
29      if ((key >= "A") and (key <= "D")) then
30         strength=key                   -- set strength
31      else
32         if ((key >= "0") and (key <= "9")) then
33            relayInSec=tonumber(key)     -- set cooking second
34         else
35            if (key == "*") then         -- start cooking
36               tmr.stop(0)              -- disable keypad scan algorithm
37               cooking(relayInSec)      -- start cooking
38               tmr.start(0)             -- enable keypad scan algorithm
39            end
40         end
41      end
42      print("sec=",relayInSec,"s=",strength,"k=",key)
43   end
44   function cooking(sec)
45      gpio.write(relayPin,gpio.HIGH)     -- turn on heating (relay and turnner)
46      for i=1,relayInSec do              -- count down
47         tmr.delay(1000000)
48         print(i,"second")
49      end
50      gpio.write(relayPin,gpio.LOW)      -- turn off heating
51   end
```

圖 11-6

(2)　利用 Snap4NodeMCU 平台實做出微波爐程式設計，如圖 11-7(a)、(b)、(c)和(d)。

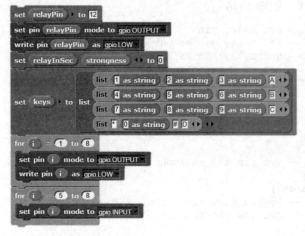

圖 11-7(a)

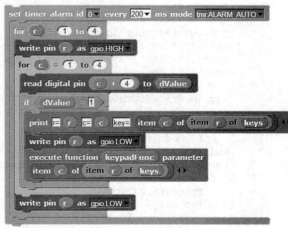

圖 11-7(b)

圖 11-7(c)

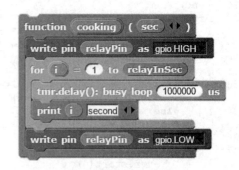

圖 11-7(d)

主題 A：NodeMCU 實作 MQTT 發佈以及訂閱訊息

1. 題目：兩人一組，撰寫 Lua 程式實驗下述兩點：

   (1) 組員 A 在 iot.eclipse.org 的 Broker 中註冊並訂閱一個"topicA"的 Topic；組員 B 則是針對該 Topic 發佈一個"Hello, I am B"的訊息。

   (2) 組員 B 在 iot.eclipse.org 的 Broker 中註冊並訂閱一個"topicB"的 Topic；組員 A 則是針對該 Topic 發佈一個" Hello B"的訊息。

2. 實驗目標：讓 NodeMCU A/B 可以使用 MQTT 傳送訊息至 NodeMCU B/A，並且讓 NodeMCU A/B 使用 MQTT 接收來自 NodeMCU B/A 的訊息。

3. 實驗步驟：

   (1) 準備兩個 NodeMCU，一個當 A 另一個當 B。同時開兩個 ESPlorer 分別和兩個 NodeMCU 連線，在 ESPlorer 的編輯畫面撰寫 Wi-Fi 連線的程式，如圖 12-1(a)。其中 wifi.sta.config(ssid, password)函式用來設定 Wi-Fi AP 的 ssid 以及密碼，所以使用者需要輸入自己 AP 的 ssid 以及密碼。也可以使用 Snap4NodeMCU 加以實現，如圖 12-1(b)。注意：如果有本中心的智慧閘道器，可以直接使用自動安全連線，如圖 12-1(c)。

```
1   wifi.setmode(wifi.STATION)                  -- set ESP8266 in STATION mode
2   wifi.sta.config("AP-SSID","AP-password")    -- set SSID and password to connect to AP
3 ┌ tmr.alarm(0,1000,tmr.ALARM_AUTO,function()  -- repeat every 1 second
4     ip,netmask,gateway=wifi.sta.getip()       -- read the ip info
5     print("ip=",ip)                           -- print out ip address
6 ┌   if (not (ip == nil)) then                 -- if ip is not nill then connect to AP
7       tmr.stop(0)                             -- stop timer 0
8     end
9 └ end)
```

圖 12-1(a)

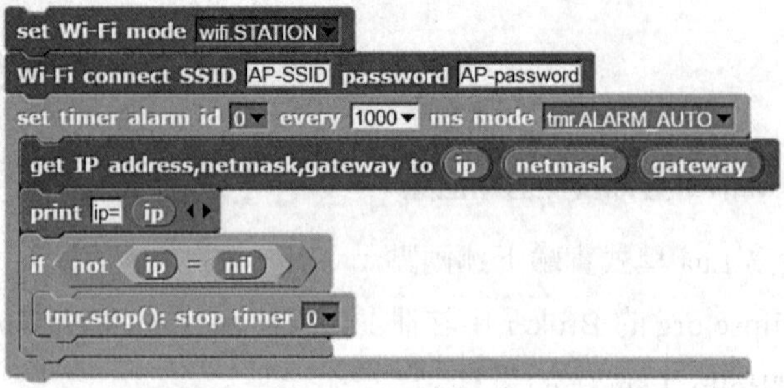

圖 12-1(b)

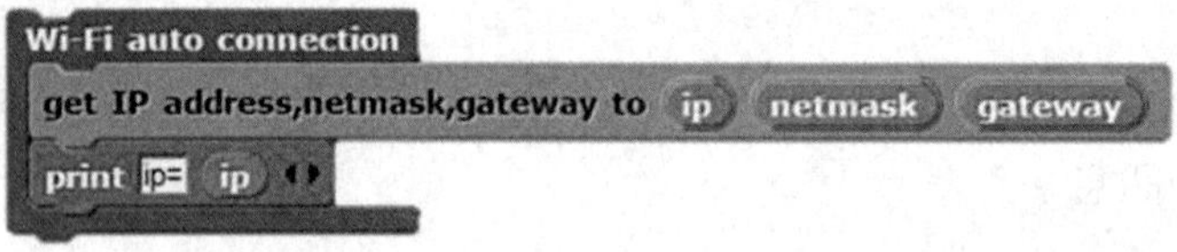


```lua
require("C4lab_AutoConnect")

function __start()
  ip,netmask,gateway=wifi.sta.getip()
  print("ip=",ip)
  -- put MQTT code here
end
```

圖 12-1(c)

(2) 連上指定的 Wi-Fi AP 後，印出 NodeMCU 獲得 IP 的訊息，如圖 12-2。左邊爲 A 的 IP，右邊爲 B 的 IP。

A

```
> dofile("wifi.lua");
> nil
nil
192.168.1.212   255.255.255.0   192.168.1.1
```

B

```
> dofile("wifi.lua");
> nil
nil
nil
192.168.1.170   255.255.255.0   192.168.1.1
```

圖 12-2

(3) 當已經連上 AP 之後，A 當訂閱著(Subscriber)角色，在 iot.eclipse.org 的 Broker 中註冊並訂閱一個"topicA"的 Topic，當有訊息發佈到"topicA"時，A 就會收到，如圖 12-3(a)。也可以使用 Snap4NodeMCU 加以實現，如圖 12-3(b)。注意：如果有本中心的智慧閘道器，可以直接使用自動安全連線，如圖 12-1(c)。

```lua
__mqtt = mqtt.Client("clientId_A",120,"username","password")    -- create a MQTT object
__mqtt:connect("iot.eclipse.org",1883,0,0, function(__mqtt)     -- connect to MQTT broker (iot.eclipse.org)
  print("connection ok!")                                       -- print out conn OK if success
  __mqtt:subscribe("topicA",0,function(__mqtt)                  -- sub topicA from iot.eclipse.org
    print("sub ok!")                                            -- print out subscribe OK if success
    __mqtt:on("message", function(__mqtt,topic,data)            -- register callback function
      print("topic=",topic,"message=",data)                     -- print out received MQTT message
    end)
  end)
end)
```

圖 12-3(a)

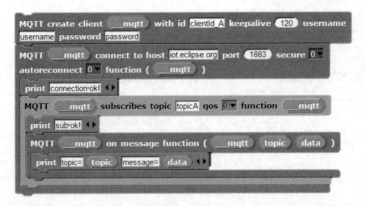

圖 12-3(b)

(4) 當已經連上 AP 之後，B 當發布者(Publisher)角色，在 iot.eclipse.org 的 Broker 中發佈訊息到"topicA"，如圖 12-4(a)。也可以使用 Snap4NodeMCU 加以實現，如圖 12-4(b)。注意：如果有本中心的智慧閘道器，可以直接使用自動安全連線，如圖 12-1(c)。

```
1   __mqtt = mqtt.Client("clientId_B",120,"username","password")      -- create a MQTT object
2   __mqtt:connect("iot.eclipse.org",1883,0,0, function(__mqtt)       -- connect to MQTT broker (iot.eclipse.org)
3     print("connection ok!")                                         -- print out conn OK if success
4     __mqtt:publish("topicA","Hello, I am B",0,0, function(__mqtt)   -- publish "Hello, I am B" message to topicA
5       print("message sent!")                                        -- print out publish OK if success
6     end)
7   end)
```

圖 12-4(a)

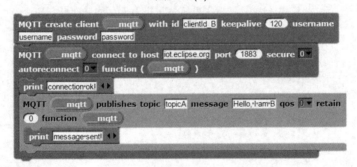

圖 12-4(b)

(5) 當已經連上 AP 之後，B 當訂閱著(Subscriber)角色，在 iot.eclipse.org 的 Broker 中註冊並訂閱一個"topicB"的 Topic，因此當有訊息發佈到"topicB"時，B 就會收到，如圖 12-5。

```
1   --clientid cannot be the same with publisher
2   m = mqtt.Client("clientid_B", 120, nil, nil)
3   m:connect("iot.eclipse.org", 1883, 0, function(m)
4     print("clientid B connect to MQTT Broker success")
5     m:subscribe("topicB",0, function(m)
6       print("clientid_B subscribe topicB success")
7     end)
8   end)
9
10  m:on("message", function(m, topic, data)
11    print("Topic ["..  topic .. "]:" )
12    if data ~= nil then
13      print(data)
14    end
15  end)
```

```
1   m = mqtt.Client()"clientid_A", 120, nil, nil)|
2   m:connect("iot.eclipse.org", 1883, 0, function(m)
3     print("clientid_A connect to MQTT Broker success")
4   end)
5
6   m:on("connect", function(m)
7     print ("clientid A connected")
8     msg="Hello B"
9     m:publish("topicB",msg,0,0, function(m)
10      print("\n clientid_A sent \""..msg.."\" success")
11    end)
12  end)
```

圖 12-5   圖 12-6

(6) 當已經連上 AP 之後，A 當發布者(Publisher)角色，在 iot.eclipse.org 的 Broker 中發佈訊息到"topicB"，如圖 12-6。

4. 實驗成果：

(1) 組員 A 在 iot.eclipse.org 的 Broker 中註冊並訂閱一個"topicA"的 Topic；組員 B 則是針對該 Topic 發佈一個"Hello, I am B"的訊息，終端機顯示畫面如圖 12-7 至圖 12-9。

```
file.remove("Subscribe.lua");
> file.open("Subscribe.lua","w+");
> w = file.writeline
> w([[--clientid cannot be the same with publisher]]);
> w([[m = mqtt.Client("clientid_A", 120, nil, nil)]]);
> w([[m:connect("iot.eclipse.org", 1883, 0, function(m) ]]);
> w([[  print("connect to MQTT Broker success")]]);
> w([[  m:subscribe("topicA",0, function(m) ]]);
> w([[    print("subscribe success") ]]);
> w([[   end) ]]);
> w([[end)]]);
> w([[]]);
> w([[m:on("message", function(m, topic, data) ]]);
> w([[  print("Topic ["..  topic .. "]:" ) ]]);
> w([[  if data ~= nil then]]);
> w([[    print(data)]]);
> w([[  end]]);
> w([[end)]]);
> file.close();
> dofile("Subscribe.lua");       A的終端機畫面
> connect to MQTT Broker success
subscribe success
Topic [topicA]:
mood test
```

圖 12-7

```
file.remove("Publish.lua");
> file.open("Publish.lua","w+");
> w = file.writeline
> w([[m = mqtt.Client("clientid_B", 120, nil, nil)]]);
> w([[m:connect("iot.eclipse.org", 1883, 0, function(m) ]]);
> w([[ print("Publisher connect...")]]);
> w([[end)]]);
> w([[]]);
> w([[m:on("connect", function(m) ]]);
> w([[ print ("connect") ]]);
> w([[ msg="Hello, I am B"]]);
> w([[ m:publish("topicA",msg,0,0, function(m) ]]);
> w([[   print("\n clientid_B sent \""..msg..\" success") ]]);
> w([[ end)]]);
> w([[end)]]);
> file.close();
> dofile("Publish.lua");
> connect

clientid_B sent "Hello, I am B" success
```

B的終端機畫面

```
> dofile("Subscribe.lua");
> connect to MQTT Broker success
subscribe success
Topic [topicA]:
mood test
Topic [topicA]:
Hello, I am B
```

B發佈訊息時
A的終端機
畫面

圖 12-8

圖 12-9

(2) 組員 B 在 iot.eclipse.org 的 Broker 中註冊並訂閱一個"topicB"的 Topic；組員 A 則是針對該 Topic 發佈一個" Hello B"的訊息，終端機顯示畫面如圖 12-10。

```
file.remove("A_Publish.lua");
> file.open("A_Publish.lua","w+");
> w = file.writeline
> w([[m = mqtt.Client("clientid_A", 120, nil, nil)]]);
> w([[m:connect("iot.eclipse.org", 1883, 0, function(m) ]]);
> w([[ print("clientid_A connect to MQTT Broker success")]]);
> w([[end)]]);
> w([[]]);
> w([[m:on("connect", function(m) ]]);
> w([[ print ("clientid_A connected") ]]);
> w([[ msg="Hello B"]]);
> w([[ m:publish("topicB",msg,0,0, function(m) ]]);
> w([[   print("\n clientid_A sent \""..msg.."\" success") ]]);
> w([[ end)]]);
> w([[end)]]);
> file.close();
> dofile("A_Publish.lua");
> clientid_A connected

 clientid_A sent "Hello B" success
```

```
file.remove("B_Subscribe.lua");
> file.open("B_Subscribe.lua","w+");
> w = file.writeline
> w([[--clientid cannot be the same with publisher]]);
> w([[m = mqtt.Client("clientid_B", 120, nil, nil)]]);
> w([[m:connect("iot.eclipse.org", 1883, 0, function(m) ]]);
> w([[ print("clientid_B connect to MQTT Broker success")]]);
> w([[ m:subscribe("topicB",0, function(m) ]]);
> w([[   print("clientid_B subscribe topicB success") ]]);
> w([[ end) ]]);
> w([[end)]]);
> w([[]]);
> w([[m:on("message", function(m, topic, data) ]]);
> w([[ print("Topic ["..  topic .. "]:" ) ]]);
> w([[ if data ~= nil then]]);
> w([[   print(data)]]);
> w([[ end]]);
> w([[end)]]);
> file.close();
> dofile("B_Subscribe.lua");
> clientid_B connect to MQTT Broker success
clientid_B subscribe topicB success
Topic [topicB]:
Hello B
```

圖 12-10

135

主題 B：NodeMCU 實作 MQTT 結合 LED 燈控制

1. 題目：兩人一組，撰寫 Lua 程式實驗下述兩點：

   (1) 組員 A 在 iot.eclipse.org 的 Broker 中偵測到組員 B 送的訊息到 Topic "topicA"，組員 A 依據訊息內容開啓或關閉他的 LED 燈。

   (2) 組員 B 在 iot.eclipse.org 的 Broker 中偵測到組員 A 送了訊息到 Topic "topicB"，組員 B 依據訊息內容開啓或關閉他的 LED 燈。。

2. 實驗目標：讓 NodeMCU A/B 可以使用 MQTT 傳送訊息至 NodeMCU B/A，並且讓 NodeMCU A/B 使用 MQTT 接收來自 NodeMCU B/A 的訊息；若接收訊息爲 on，則 LED 燈亮，訊息爲 off，則 LED 燈暗。

3. 實驗步驟：

   (1) 延續主題 A，先進行 Wi-Fi 連線獲得 IP 的訊息。

   (2) A 當訂閱著(Subscriber)角色，在 iot.eclipse.org 的 Broker 中註冊並訂閱一個"topicA"的 Topic，並設定當收到訊息是 on 時則開啓 NodeMCU D4 的 LED 燈，當收到訊息是 off 時則關閉 NodeMCU D0 的 LED 燈，如圖 12-11(a)。也可以使用 Snap4NodeMCU 加以實現，如圖 12-11(b)。

```
1   led=4
2   gpio.mode(led,gpio.OUTPUT)
3   __mqtt = mqtt.Client("clientId_A",120,"username","password")
4   __mqtt:connect("iot.eclipse.org",1883,0,0, function(__mqtt)
5     print("connection ok!")
6     __mqtt:subscribe("topicA",0,function(__mqtt)
7       print("sub ok!")
8       __mqtt:on("message", function(__mqtt,topic,data)
9         print("topic=",topic,"message=",data)
10        if (data == "on") then
11          print("Light on")
12          gpio.write(led,gpio.HIGH)
13        end
14        if (data == "off") then
15          print("Light off")
16          gpio.write(led,gpio.LOW)
17        end
18      end)
19    end)
20  end)
```

圖 12-11(a)

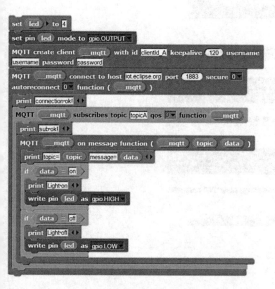

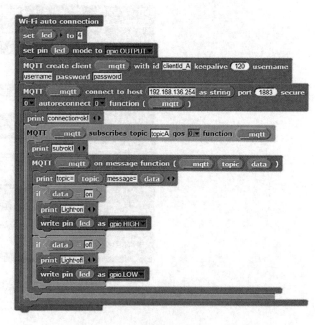

圖 12-11(b)   圖 12-11(c)

(3) B 當發布者(Publisher)角色，在 iot.eclipse.org 的 Broker 中發佈訊息到"topicA"，每三秒變更
    訊息一次，一開始發佈訊息為"on"，三秒後改為"off"，如圖 12-12(a)。也可以使用
    Snap4NodeMCU 加以實現，如圖 12-12(b)。

```
1     __mqtt = mqtt.Client("clientId_B",120,"username","password")
2     __mqtt:connect("iot.eclipse.org",1883,0,0, function(__mqtt)
3       print("connection ok!")
4       msg="on"
5       tmr.alarm(0,3000,tmr.ALARM_AUTO,function()
6         if (msg == "on") then
7           __mqtt:publish("topicA","on",0,0, function(__mqtt)
8             print("on message sent!")
9           end)
10          msg="off"
11        else
12          __mqtt:publish("topicA","off",0,0, function(__mqtt)
13            print("off message sent!")
14          end)
15          msg="on"
16        end
17      end)
18    end)
```

圖 12-12(a)

137

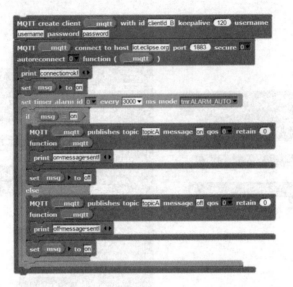

圖 12-12(b)

(4) 接著換 B 當訂閱著(Subscriber)角色，在 iot.eclipse.org 的 Broker 中註冊並訂閱一個"topicB"的 Topic，程式只要修改圖 12-11 的 A 為 B。而 A 當發布者 (Publisher)角色，在 iot.eclipse.org 的 Broker 中發佈訊息到"topicB"，也只要修改的 A 為 B、B 為 A，如圖 12-13。

```lua
1  m = mqtt.Client("clientid_A", 120, nil, nil)
2  m:connect("iot.eclipse.org", 1883, 0, function(m)
3    print("clientid_A connect to MQTT Broker success")
4  end)
5
6  msg="on"                                            A為Publisher
7  m:on("connect", function(m)
8    print ("clientid_A connected")
9    tmr.alarm(0,3000,1, function()
10     if msg=="on" then
11       m:publish("topicB",msg,0,0, function(m)
12         print("\n clientid_A sent \""..msg.."\" success")
13         msg="off"
14       end)
15     else
16       m:publish("topicB",msg,0,0, function(m)
17         print("\n clientid_A sent \""..msg.."\" success")
18         msg="on"
19       end)
20     end
21   end)
22 end)
```

```lua
1  m = mqtt.Client("clientid_B", 120, nil, nil)
2  m:connect("iot.eclipse.org", 1883, 0, function(m)
3    print("clientid_B connect to MQTT Broker success")
4    m:subscribe("topicB",0, function(m)
5      print("clientid_B subscribe topicB success")
6    end)
7  end)                                              B為Subscriber
8
9  m:on("message", function(m, topic, data)
10   print("Topic [".. topic .. "]:" )
11   if data ~= nil and data == "on"  then
12     print("Light on")
13     gpio.write(0,gpio.LOW)
14   elseif data ~= nil and data=="off" then
15     print("Light off")
16     gpio.write(0, gpio.HIGH)
17   end
18 end)
```

圖 12-13

138

4. 實驗成果：

(1) 組員 A 在 iot.eclipse.org 的 Broker 中偵測到組員 B 送的訊息到 Topic "topicA"，組員 A 依據訊息內容開啓或關閉他的 LED 燈，終端機顯示畫面以及實際操作畫面如圖 12-14。

```
> clientid_A connect to MQTT Broker success        clientid_B sent "on" success
clientid_A subscribe topicA success               clientid_B connected
Topic [topicA]:
Topic [topicA]:                                    clientid_B sent "on" success
Light on
Topic [topicA]:                                    clientid_B sent "off" success
Light off
Topic [topicA]:                                    clientid_B sent "on" success
Light on
Topic [topicA]:                                    clientid_B sent "off" success
Light off
Topic [topicA]:                                    clientid_B sent "on" success
```

圖 12-14

(2) 組員 B 在 iot.eclipse.org 的 Broker 中偵測到組員 A 送的訊息到 Topic "topicB"，組員 B 依據訊息內容開啓或關閉他的 LED 燈，終端機顯示畫面以及實際操作畫面如圖 12-15。

```
> clientid_A connected                   > clientid_B connect to MQTT Broker success
                                         clientid_B subscribe topicB success
clientid_A sent "on" success             Topic [topicB]:
                                         Topic [topicB]:
clientid_A sent "off" success            Light off
                                         Topic [topicB]:
clientid_A sent "on" success             Light on
                                         Topic [topicB]:
clientid_A sent "off" success            Light off
                                         Topic [topicB]:
clientid_A sent "on" success             Light on
```

139

圖 12-15

主題 C：NodeMCU 實作 MQTT 結合移動感測器進行 LED 燈控制

1. 題目：加入壓力感測器、超音波感測器或是移動感測器，利用 MQTT 機制，當一個物聯網設備偵測到有人的時候，另一個物聯網設備自動點亮燈光。

2. 實驗目標：如同主題 B 讓 NodeMCU A/B 可以使用 MQTT 傳送以及接收訊息；若偵測到有人時，則 LED 燈亮，反之則 LED 燈暗。

3. 實驗步驟：

    (1) 先進行 Wi-Fi 連線獲得 IP 的訊息。

    (2) 如同主題 B，在 A 當訂閱著 (Subscriber)角色時的程式碼並不需要修改，如圖 12-16。

    (3) B 當發佈者(Publisher)角色，在 iot.eclipse.org的 Broker 中發佈訊息到"topicA"，每三秒偵測一次 PIR 是否有人，當偵測有人時發佈訊息為"on"，偵測沒有人時發佈訊息為"off"，如圖 12-17。

```
1   m = mqtt.Client("clientid_A", 120, nil, nil)
2   m:connect("iot.eclipse.org", 1883, 0, function(m)
3     print("clientid_A connect to MQTT Broker success")
4     m:subscribe("topicA",0, function(m)
5       print("clientid_A subscribe topicA success")
6     end)
7   end)
8
9   m:on("message", function(m, topic, data)
10    print("Topic [".. topic .. "]:" )
11    if data ~= nil and data == "on"  then
12      print("Light on")
13      gpio.write(0,gpio.LOW)
14    elseif data ~= nil and data=="off" then
15      print("Light off")
16      gpio.write(0, gpio.HIGH)
17    end
18  end)
```

圖 12-16

```
1   m = mqtt.Client("clientid_B", 120, nil, nil)|
2   m:connect("iot.eclipse.org", 1883, 0, function(m)
3     print("clientid_B connect to MQTT Broker success")
4   end)
5
6   PIRPin=2
7   gpio.mode(PIRPin,gpio.INPUT)
8   m:on("connect", function(m)
9     print ("clientid_B connected")
10    tmr.alarm(0,3000,1, function()
11      state=gpio.read(PIRPin)
12      if state ==1 then
13          msg="on"
14        m:publish("topicA",msg,0,0, function(m)
15            print("state="..state.." clientid_B sent \""..msg.."\" success")
16        end)
17      else
18          msg="off"
19        m:publish("topicA",msg,0,0, function(m)
20            print("state="..state.." clientid_B sent \""..msg.."\" success")
21        end)
22      end
23    end)
24  end)
```

圖 12-17

4. 實驗成果：

組員 A 在 iot.eclipse.org 的 Broker 中偵測到組員 B 送的訊息到 Topic "topicA"，組員 A 依據訊息內容開啓或關閉他的 LED 燈，終端機顯示畫面以及實際操作畫面如圖 12-18 和圖 12-19。

```
> clientid_A connect to MQTT Broker success      > clientid_B connected
clientid_A subscribe topicA success              state=1 clientid_B sent "on" success
Topic [topicA]:                                  state=1 clientid_B sent "on" success
Topic [topicA]:                                  state=1 clientid_B sent "on" success
Light on                                         state=1 clientid_B sent "on" success
Topic [topicA]:          有偵測到人              state=1 clientid_B sent "on" success
                                                 state=1 clientid_B sent "on" success
                                                 state=1 clientid_B sent "on" success
---------------------------------
Topic [topicA]:                                  state=0 clientid_B sent "off" success
Light off                                        state=0 clientid_B sent "off" success
Topic [topicA]:                                  state=0 clientid_B sent "off" success
Light off                                        state=0 clientid_B sent "off" success
Topic [topicA]:          沒有偵測到人            state=0 clientid_B sent "off" success
Light off                                        state=0 clientid_B sent "off" success
Topic [topicA]:
Light off
```

圖 12-18

141

沒有偵測到人

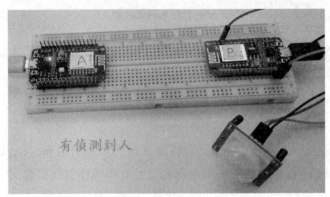

有偵測到人

圖 12-19

主題 A：使用 FritzingIDE 平台建構溫濕度感測器、LED 燈和光感測元件的電路圖

1. 題目：使用 Fritzing 建構包含溫濕度感測器、LED 燈和光感測元件的電路圖以及實際麵包板接線圖。

2. 實驗目標：利用 Fritzing 將要進行 CoAP 實驗的感測器實際接線，讓使用者實際體驗線路的建構。

3. 實驗步驟：

   (1) 開啟 Fritzing，其初始畫面如圖 13-1。

   (2) 在右方搜尋窗格輸入"NodeMCU"，將 NodeMCU 往麵包板上拖曳並放置在麵包板上，如圖 13-2。

   (3) 搜尋"RHT"，將溫濕度感測器拖曳至麵包板上，如圖 13-3。

圖 13-1

圖 13-2

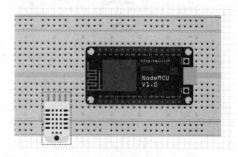

圖 13-3

   (4) 將溫濕度感測器的 VCC 接腳接到麵包板上的電源輸入接孔，GND 接腳接到麵包板上的接地，如圖 13-4。

圖 13-4

(5) 將溫濕度感測器的 Data 接腳接線到 NodeMCU 上的 D1，並將 NodeMCU 的 3V3 接到麵包板上的電源接孔，GND 接到麵包板上的接地接孔，如圖 13-5。

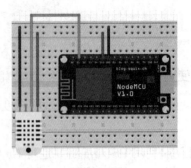

圖 13-5

(6) 點選上方的"概要圖"，並將元件以及線路擺放至較美觀的位置，並將線路佈線完成，如圖 13-6。

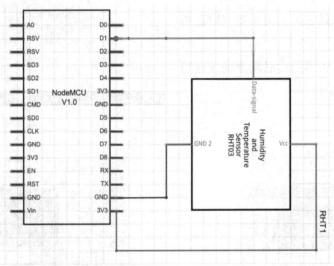

圖 13-6

(7) 搜尋"LED"和"resistor"，將 LED 和電阻拖曳至麵包板上，如圖 13-7。

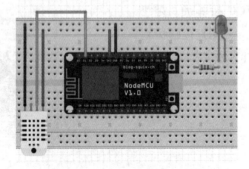

圖 13-7

(8) 將 LED 的長邊接腳接到 NodeMCU 上的 D2 接孔，短邊接腳和電阻相接，而電阻另一端接腳接到麵包板上的 GND，如圖 13-8。

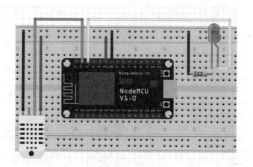

圖 13-8

(9) 搜尋"LDR"和"resistor"，將光感測元件和電阻拖曳至麵包板上，如圖 13-9。

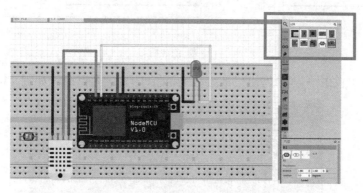

圖 13-9

(10) 將光感測元件的接腳一邊接到 NodeMCU 上的 A0 接孔，一邊接到麵包板上的 GND，並將電阻的另一端接腳接到麵包板上的電源接孔，如圖 13-10。

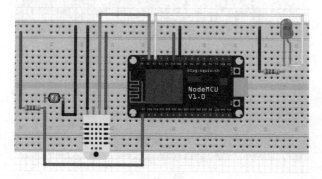

圖 13-10

(11) 點選上方的"概要圖"，將目前的電路示意圖的元件以及線路接好，如圖 13-11。

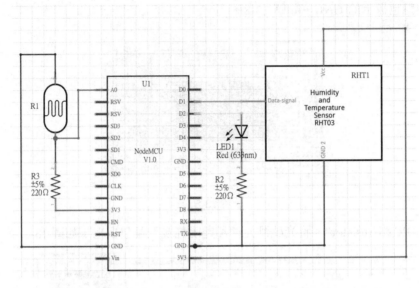

圖 13-11

(12) 點選上方的"PCB"，則會秀出目前接好的 PCB 印刷版電路圖，如圖 13-12。

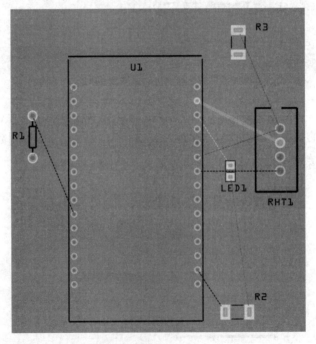

圖 13-12

(13) 接著將元件擺放至適當位置並且將線路佈線完成，如圖 13-13。

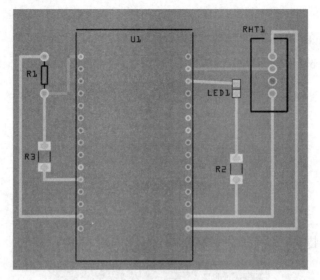

圖 13-13

(14) 實際麵包板接線可以依照由 Fritzing 所拉的電路圖來實作，如圖 13-14 和圖 13-15。

圖 13-14

圖 13-15

主題 B：溫濕度感測器 CoAP server 實作

1. 題目：撰寫 CoAP Server 的 Lua 程式讀取溫濕度感測器的溫度和濕度值，並觀察以及實驗下述三點：

(1) 下載 Google android phone "智慧控制"App。

(2) 手輕輕按住感測模組，觀察"智慧控制"App 的變化。

(3) 對著感測模組吹氣，觀察"智慧控制"App 的變化。

注意：本實驗必須購買本中心的智慧閘道器，才有提供自動安全連線和 CoAP 功能。

2. 實驗目標：利用 NodeMCU 進行 CoAP 讀取溫濕度感測器訊息傳遞實驗。

3. 實驗步驟：

(1) 下載 Google android phone "智慧控制"App。

    (a) 利用 Android 手機至 Play 商店搜尋"智慧控制"，如圖 13-16。

圖 13-16

    (b) 點選後安裝本中心開發的"智慧控制" App，如圖 13-17。

圖 13-17

(2) 溫濕度感測器接線圖

    利用麵包板將溫濕度感測器依照接線圖將線路接好，如圖 13-18。

圖 13-18

(3) 在 ESPlorer 的編輯畫面撰寫控制溫濕度感測器的程式,如圖 13-19(a)。也可以使用 Snap4NodeMCU 加以實現,如圖 13-19(b)。

```lua
1    require("C4lab_AutoConnect")
2    tmr.alarm(0,1000,tmr.ALARM_AUTO,function()
3      status,temp,humi,temp_decimial,humi_decimial=dht.read(7)
4      print("t=",temp,"h=",humi)
5    end)
6    deviceInfo = {
7      flags = 0x00000001,
8      classID = 0,
9      deviceID = 120
10   }
11
12   function typeID(p)
13     if p~=nil then
14       deviceInfo.deviceID = p
15     end
16     return deviceInfo.deviceID
17   end
18
19   function flags(p)
20     if p~=nil then
21       deviceInfo.flags = p
22     end
23     return deviceInfo.flags
24   end
25
26   __cv = require "C4lab_CoAP210-ob"
27   __cv:startCoapServer(5683)
28   __cv:addActuator("/typeID","typeID")
29   __cv:addActuator("/flags","flags")
30   __cv:addSensor("/121/0/1","temp")
31   __cv:addSensor("/121/0/2","humi")
32   __cv:postResource()
```

圖 13-19(a)

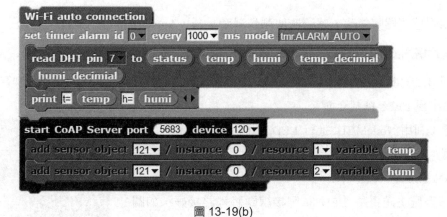

圖 13-19(b)

149

(4) 執行"自動連線"程式

開啟智慧閘道器，並點選本中心專利"自動連線"程式，讓 NodeMCU 連上智慧閘道器。

(5) 開啟"智慧控制"App

在執行 App 程式前必須先開啟手機 Wi-Fi 功能，接著點選手機應用程式畫面上的"智慧控制"App，如圖 13-20。開啟後即會自動連線至 Gateway，搜尋可用的 IoT Gateway 設備，如圖 13-21。

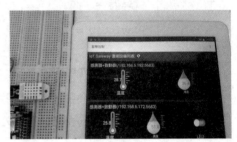

圖 13-20

圖 13-21

4. 實驗成果：

(1) 當手指壓住感測器時，體溫使得感測器溫度上升，App 畫面如圖 13-22。

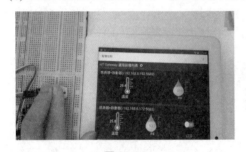

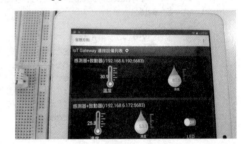

圖 13-22

圖 13-23

(2) 以口對感測器吹氣時，感測器偵測到濕度上升，App 畫面如圖 13-23。

主題 C：溫濕度感測器結合 LED 之 CoAP server 實作

1. 題目：加入 LEDs 模組，撰寫 CoAP Server 的 Lua 程式去開啟或關閉 LED，並觀察"智慧控制"App 是否自動出現 LEDs 控制介面。

2. 實驗目標：利用 NodeMCU 進行 LED 之 CoAP 實驗。

3. 實驗步驟：

(1) LED 接線圖

接續上個主題再將 LED 依照接線圖將線路接好，如圖 13-24。

圖 13-24

(2) 在 ESPlorer 的編輯畫面撰寫控制 LED 的程式，如圖 13-25(a)。也可以使用 Snap4NodeMCU 加以實現，如圖 13-25(b)。

```lua
1     require("C4lab_AutoConnect")
2     led=0
3     gpio.mode(4,gpio.OUTPUT)
4     gpio.write(4,gpio.HIGH)
5     tmr.alarm(0,1000,tmr.ALARM_AUTO,function()
6        status,temp,humi,temp_decimial,humi_decimial=dht.read(7)
7        print("t=",temp,"h=",humi)
8     end)
9     function setLED(payload)
10       if (payload == "1") then
11          led=1
12          gpio.write(4,gpio.LOW)
13       end
14       if (payload == "0") then
15          led=0
16          gpio.write(4,gpio.HIGH)
17       end
18       return led
19    end
20    deviceInfo = {
21       flags = 0x00000001,
22       classID = 0,
23       deviceID = 120
24    }
25
26    function typeID(p)
27       if p~=nil then
28          deviceInfo.deviceID = p
29       end
30       return deviceInfo.deviceID
31    end
32
33    function flags(p)
34       if p~=nil then
35          deviceInfo.flags = p
36       end
37       return deviceInfo.flags
38    end
39
40    __cv = require "C4lab_CoAP210-ob"
41    __cv:startCoapServer(5683)
42    __cv:addActuator("/typeID","typeID")
43    __cv:addActuator("/flags","flags")
44    __cv:addSensor("/121/0/1","temp")
45    __cv:addSensor("/121/0/2","humi")
46    __cv:addActuator("/122/0/201","setLED")
47    __cv:postResource()
```

圖 13-25(a)

圖 13-25(b)

(3) 執行"自動連線"程式

點選本中心專利"自動連線"程式,讓 NodeMCU 連上 Gateway。

(4) 開啟"智慧控制"App

點選手機應用程式畫面上的"智慧控制"App,開啟後即會自動連線至 Gateway,搜尋可用的 IoT Gateway 設備。

4. 實驗成果:

(1) "智慧控制"App 會自動出現 LEDs 控制介面,如圖 13-26。

圖 13-26

圖 13-27

(2) 利用控制介面將 LED 打開時麵包板上的 LED 燈就會亮起來,實際畫面如圖 13-27。

主題 D：溫濕度感測器結合 LED 及光感測器之 CoAP server 實作

1. 題目：加入光感測器元件，撰寫 CoAP Server 的 Lua 程式去讀取光感測器元件明亮度，並觀察以及實驗下述兩點：

   (1) 當手不蓋/半蓋/蓋住光感測器，觀察"智慧控制"App 的變化。

   (2) 使用手機"手電筒"App 照射感測器，觀察"智慧控制"App 的變化。

2. 實驗目標：利用 NodeMCU 進行光感測器元件之 CoAP 實驗。

3. 實驗步驟：

   (1) 光感測器元件接線圖
   接續上個主題再將光感測器元件依照
   接線圖將線路接好，如圖 13-28。

圖 13-28

   (2) 在 ESPlorer 的編輯
   畫面撰寫光感測器
   元件的程式，如圖
   13-29(a)。

```lua
1   require("C4lab_AutoConnect")
2   led=0
3   gpio.mode(4,gpio.OUTPUT)
4   gpio.write(4,gpio.HIGH)
5   tmr.alarm(0,1000,tmr.ALARM_AUTO,function()
6     status,temp,humi,temp_decimial,humi_decimial=dht.read(7)
7     aValue=adc.read(0)
8     print("t=",temp,"h=",humi,"l=",aValue)
9   end)
10  function setLED(payload)
11    if (payload == "1") then
12      led=1
13      gpio.write(4,gpio.LOW)
14    end
15    if (payload == "0") then
16      led=0
17      gpio.write(4,gpio.HIGH)
18    end
19    return led
20  end
21  deviceInfo = {
22    flags = 0x00000001,
23    classID = 0,
24    deviceID = 120
25  }
26
27  function typeID(p)
28    if p~=nil then
29      deviceInfo.deviceID = p
30    end
31    return deviceInfo.deviceID
32  end
33
34  function flags(p)
35    if p~=nil then
36      deviceInfo.flags = p
37    end
38    return deviceInfo.flags
39  end
40
41  _cv = require "C4lab_CoAP210-cb"
42  _cv:startCoapServer(5683)
43  _cv:addActuator("/typeID","typeID")
44  _cv:addActuator("/flags","flags")
45  _cv:addSensor("/121/0/1","temp")
46  _cv:addSensor("/121/0/2","humi")
47  _cv:addActuator("/122/0/201","setLED")
48  _cv:postResource()
```

圖 13-29(a)

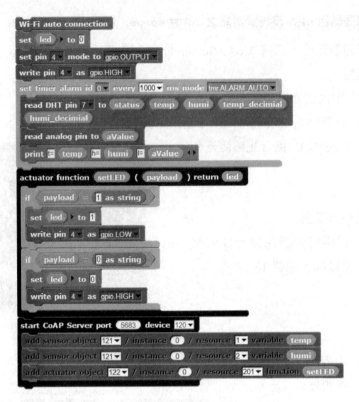

圖 13-29(b)

(3) 執行"自動連線"程式

　　點選本中心專利"自動連線"程式，讓 NodeMCU 連上 Gateway。

(4) 開啓"智慧控制"App

　　點選手機應用程式畫面上的"智慧控制"App，開啓後即會自動連線至 Gateway，搜尋可用的 IoT Gateway 設備。

4. 實驗成果：

(1) 在"智慧控制"App 加入光感測器元件後且手不蓋的畫面如圖 13-30。

圖 13-30

(2) 手半蓋後 App 畫面如圖 13-31，可以發現亮度值由 119 升到 274。

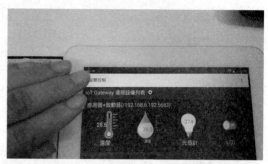

圖 13-31

(3) 手全蓋後 App 畫面如圖 13-32，可以發現亮度值已經升到 661。

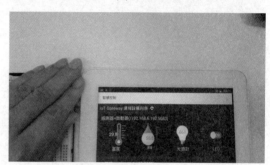

圖 13-32

(4) 利用"手電筒"照射感測器，App 畫面如圖 13-33，可以發現亮度值已經下降到 71。
原因是光感測器元件是透過分壓原理得知亮度，因此當亮度越亮讀到的值越小，當亮度越暗時讀到的值越大。

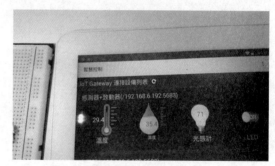

圖 13-33